T0135584

A mathematical modeling framework to simulate and analyze cell type transitions

Von der Fakultät Konstruktions-, Produktions- und Fahrzeugtechnik und dem Suttgart Research Centre for Simulation Technology der Universität Stuttgart zur Erlangung der Würde eines Doktors der Ingenieurswissenschaften (Dr.-Ing.) genehmigte Abhandlung

Vorgelegt von

Daniella Schittler

aus München

Hauptberichter: Prof. Dr.-Ing. Frank Allgöwer
Mitberichter: Prof. Dr. Rob J. De Boer
Jun.-Prof. Dr.-Ing. Steffen Waldherr

Tag der mündlichen Prüfung: 02.02.2015

Institut für Systemtheorie und Regelungstechnik
Universität Stuttgart
2015

Bibliografische Information der Deutschen Nationalbibliothek

Die Deutsche Nationalbibliothek verzeichnet diese Publikation in der
Deutschen Nationalbibliografie; detaillierte bibliografische Daten sind
im Internet über http://dnb.d-nb.de abrufbar.

D 93

ISBN 978-3-8325-3935-1

Logos Verlag Berlin GmbH
Comeniushof, Gubener Str. 47,
10243 Berlin
Tel.: +49 (0)30 42 85 10 90
Fax: +49 (0)30 42 85 10 92
INTERNET: http://www.logos-verlag.de

Acknowledgements

There are many people that supported me and my work, probably many more than can be covered here. This thesis presents the results of my time as a research assistant at the Institute for Systems Theory and Automatic Control (IST) at the University of Stuttgart from 2009 to 2014. At the same time, I was a PhD student at the Graduate School Simulation Technology (SimTech) at the University of Stuttgart and a fellow of The MathWorks Foundation in Science and Engineering. For three months, I was also a visiting researcher at the group for Theoretical Biology and Bioinformatics at the University of Utrecht.

First of all, I thank my supervisor Frank Allgöwer for providing plenty of scientific freedom, a very fruitful research environment, and great opportunities to visit numerous conferences with international contacts. I'd like to wholeheartedly thank Rob J. De Boer, my second supervisor, for regular inspiring discussions, invaluable positive feedback on my work, and the great hospitality in Holland. Furthermore, I'd like to greatly thank Steffen Waldherr, who was like an inofficial mentor throughout my time as a PhD student and also the third member of my thesis committee, for motivating feedback and discussions, guidance and advice.

I am grateful for all the wonderful colleagues I had at the IST, in SimTech, and in various interdisciplinary projects. First, I want to thank Jan Hasenauer for innumerable stimulating discussions, as an extraordinarily valuable colleague, co-author, as well as friend. I also want to thank my further co-authors Christian Breindl and Taouba Jouini for the great work together. Another special thanks goes to Karsten Kuritz, Jan Hasenauer, Christian Breindl, and Shen Zeng for proofreading this thesis. A big thanks goes to Beate Spinner, Claudia Surau, Claudia Vetter, Sabine Balschat, and Barbara Teutsch, for unpayable assistance in not only administrative regards.

Besides pursuing research and teaching lectures, there were also gorgeous hikes, bike tours, dinners, etc.. Thanks to all of you – Georg, Gregor, Beate, Caterina, Simon, Jingbo, Shen, Rainer, Eva, Dirke, Patrick, Christian, Jan, Rob and the group in Utrecht, and many more – for watering my tomatoes and chili plants during my absence, for affiliating me to the Söllerhaus cycling team, together conquering the Aichtal, jointly mastering steep hikes and some Gipfelbier (Kleinwalsertal) or completely flat vaksgroepsuitjes and pannenkoeken (Holland), and for company in happy and unhappy times. Many of you became true friends, and I'm very much looking forward to the future time together with you. I'd like to give a big thanks also to my dear friends outside of the university life, especially Frauke, Lucy, and Tina. This work could never have been accomplished without your unquestioning support. This page is too small to list all of you and to express my gratitude for all the wonderful times we had and will have.

Last but by no means least, I want to thank the most important people in my life: My parents Irmgard and Manfred, my sister Carina, and my husband Uwe, for always believing in me and being there for me. You know that I would not be where I am without you, and I'm unspeakably happy to have you with me.

Daniella Schittler, Stuttgart, February 2015

To all those who have always been there for me.

Contents

Index of notation

Acronyms

Acronym	Description
BrdU	Bromodeoxyuridine
CFDA-SE	Carboxyfluorescein diacetate succinimidyl ester
CFSE	Carboxyfluorescein succinimidyl ester
DNA	Deoxyribonucleic acid
DC	division- and cell type-structured
DCL	division-, cell type-, and label-structured
DsC	division-sequence- and cell type-structured
DsCL	division-sequence-, cell type-, and label-structured
GRN	gene regulatory network
mRNA	messenger Ribonucleic acid
MSC	mesenchymal stem cell
ODE	ordinary differential equation
PBE	population balance equation
PDE	partial differential equation
SDE	stochastic differential equation
TR	transcriptional regulator

Mathematics and statistics

Random variables will be denoted by upper case letters, for example, X. For the probability density of such a random variable, the corresponding lower case letter will be used, for example, x. The probability density of the random variable X to have the value x is then denoted by $p(x)$, with $\int_{\mathbb{R}} p(x)dx = 1$.

Symbol	Description
$|\mathcal{A}|$	cardinality of set $\mathcal{A}$
$||v||$	1-norm of vector v
0^T	zero vector, $[0, \dots, 0]^T$
$\mathbf{e}_j^T$	unit row vector with j-th entry 1, all other entries 0
$\dot{x}$	time derivative of x, also $\frac{dx}{dt}$
$\frac{\partial(\cdot)}{\partial x}|_{x^*}$	partial derivative, evaluated at the value $x = x^*$
$\mathrm{bino}(k|n, p)$	binomial distribution with n trials and probability p
$\mathbb{C}$	complex numbers
$\mathbb{E}(X)$	expected value of the random variable X
e^x	exponential function
I_N	identity matrix in $\mathbb{R}^{N \times N}$
$\mathbf{j}$	imaginary unit, $\sqrt{-1}$
$\log(x)$	natural logarithm
$\log \mathcal{N}(x|\mu, \sigma^2)$	lognormal distribution
M^k	with M being a set: $M \times \dots \times M$, k-times
$\mathcal{N}(x|\mu, \sigma^2)$	normal distribution
$\mathbb{N}$	natural numbers
$\mathbb{N}_0$	natural numbers including zero
$p(x)$	probability density of the random variable x
$p(x|y)$	conditional probability density of x, given y
$P(x)$	probability mass of x
$\mathcal{R}(s)$	real part of complex number s
$\mathbb{R}$	real numbers
$\mathbb{R}_+$	nonnegative real numbers, $[0, \infty)$
$\mathbb{R}_{++}$	positive real numbers, $(0, \infty)$
Ω_j	attractor basin of steady state $x^{*,j}$

Gene regulatory network models

Symbol	Description
$z(t)$, $x(t)$	vector of state variables of deterministic single cell model, at time t
$u(t)$	vector of inputs of deterministic single cell model, at time t
X_t	vector of state variables of stochastic single cell model, at time t
U_t	vector of inputs of stochastic single cell model, at time t
$f(X_t, U_t)$	drift term
$g(X_t, U_t)$	diffusion term
$G_f(\lambda)$, $G_F(\lambda)$	transfer matrix of low-/high-dimensional open-loop system
$J_f^{(r)}$, $J_F^{(r)}$	Jacobian of low-/high-dimensional system at r-th steady state
$J_{i,\nu}^=$, $J_{i,\nu}^{\neq}$	index vectors of interactions with same/opposite sign
$\lambda_{f,v}^{(r)}$, $\lambda_{F,v}^{(r)}$	v-th eigenvalue of Jacobian at r-th steady state
$\mathcal{M}_j$	interaction module of master gene j
S_a, S_A	interaction sign matrix of low-/high-dimensional system
W_t	Wiener process

Cell population models

Symbol	Description
$\alpha_i^j(t)$	division rate of cells of division number i and cell type j
$\beta_i^j(t)$	death rate of cells of division number i and cell type j
γ	label dilution factor
$\delta_i^{j_2 j_1}(t)$	cell type transition probabilities from cell type j_1 to cell type j_2
i_k	number of division undergone in the k-the time interval
L	number of labeled strands
$\nu(x, t)$	label decay rate
$m(x|t)$	number density of cells with label concentration x, at time t
$\hat{m}_S(x|t)$	number density of cells with label concentration x considering only cells of division numbers with $i_k \leq S - 1$, at time t
$n(x, \mathbf{i}, j|t)$	number density of cells with label concentration x, division number $\mathbf{i}$, cell type j, at time t
$n(y, \mathbf{i}, j|t)$	number density of cells with measured label intensity x, division number $\mathbf{i}$, and cell type j, at time t
$n(z|t)$	number density of cells with gene expression level z, at time t
$N(\mathbf{i}, j|t)$	number of cells with division number $\mathbf{i}$ and cell type j, at time t
$p(x|\mathbf{i}, j, t)$	probability density that a cell of division number $\mathbf{i}$ and cell type j at time t has label concentration x
$p_{\text{inhe}}(x|\mathbf{i} - \mathbf{e}_k^T, j, t)$	label inheritance distribution
$p_{\text{lab}}(x|t)$	effective label uptake distribution
$w_{\mathbf{i}}^{j_2 j_1}(t)$	transition-upon-division probabilities from cell type j_1 to cell type j_2
X	label concentration
Y	measured label intensity

INDEX OF NOTATION

Abstract

Living organisms are constituted by a multitude of distinct cell types, and as a crucial feature an organism's cells may regularly change their cell type. The understanding of changes in cell types, referred to as cell type transitions, is fundamental to advance fields such as stem cell research, immunology, and cancer therapies. To quantitatively comprehend and control cell type transitions, mathematical modeling and analysis methods are indispensable. Such approaches of mathematical models, analysis, and simulations are the subject of mathematical and systems biology.

This thesis provides a mathematical modeling framework to simulate and analyze cell type transitions. The novel methodological approaches and models presented here address diverse levels of detail which are essential in this context: Gene regulatory network models represent the cell type-determining gene expression dynamics, and are considered here from core motifs to a highly detailed level. The entirety of many cells, each with an individual gene regulatory network, exhibits population dynamics which are described by cell population models, capturing cell types, division numbers, as well as properties of commonly applied labeling techniques.

For a typical cell type transition process, a gene switch model is systematically developed and analyzed, yielding a model for the deterministic as well as the stochastic dynamics. This comprehensive approach exemplifies how the combination of analytical and simulation-based studies can elucidate the gene regulatory mechanisms of cell type transitions. Along with the obtained switch motif, it may serve as a blueprint for modeling gene switch dynamics in similar cell type transition processes. Furthermore, a novel construction method for high-dimensional gene regulatory network models is introduced in this thesis, which allows to derive detailed gene regulatory network models from low-dimensional core motif models. The detailed model obtained by this method conserves the core motif's multistability properties. The proposed construction procedure thereby allows to transfer results from generic low-dimensional to realistic high-dimensional gene regulatory network models.

To capture the cell type dynamics on a population level, a generalized model class is proposed that accounts for multiple cell types, division numbers, and the full label distribution. Within this new model class, analysis and solution methods are developed on a general level and independent of a specific application. Specific cell population models are then derived for two common labeling techniques. These models not only can be solved efficiently, but also allow to exploit the full information contained in the data. For tailoring models to DNA labeling techniques, a mechanistic and mathematically rigorous description for the segregation of labeled chromosomes upon cell division is developed. As demonstrated in exemplary studies on labeling data, the cell population models presented in this work unify diverse biologically relevant aspects in one model, and allow more exact and more reliable estimates of proliferation- and cell type-relevant parameters.

The modeling and analysis methods presented here connect formerly isolated approaches, such as core motif models of gene switches, detailed gene regulatory network models, cell type-structured population models, as well as cell types, division numbers and label concentrations in population models. This thesis thereby contributes to a holistic framework for the quantitative understanding of cell type transitions.

Deutsche Kurzfassung

Motivation und Fragestellung

Die Systembiologie versteht sich als eine interdisziplinäre Wissenschaft, die es sich zum Ziel gesetzt hat, mittels mathematischer Modelle biologische Prozesse quantitativ zu erfassen und nachzuvollziehen. Solche Modelle können zum Beispiel durch analytische und numerische Methoden untersucht und zur Vorhersage des Verhaltens des biologischen Systems benutzt werden. Typische Resultate betreffen Aussagen über die Stabilität von Zuständen, die Bestimmung von Wertebereichen mit qualitativ unterschiedlichem Verhalten, sowie Simulationen unterschiedlicher Szenarien. Entscheidende Vorteile der mathematischen Modellierung gegenüber experimentellen Herangehensweisen sind der meist deutlich geringere Zeit- und Kostenaufwand, sowie die Untersuchbarkeit beliebiger Szenarien, die in realen Experimenten eventuell gar nicht umsetzbar sind.

Von großem biologischen und systembiologischen Interesse ist die Untersuchung von Zelltypen, und damit verbunden von Übergängen zwischen verschiedenen Zelltypen. Der menschliche Körper besteht zum Beispiel aus etwa 200 verschiedenen Zelltypen, die jeweils eine ganz spezielle Funktion erfüllen. Übergänge zwischen Zelltypen finden statt, wenn etwa unspezialisierte Stammzellen zu spezialisierten Zellen heranreifen, wenn gesunde Zellen zu krankhaften Zellen mutieren, oder umgekehrt schädliche Zellen durch medizinische Behandlung in harmlose Zellen überführt werden können. Das quantitative Verständnis solcher Zelltypübergänge ist entscheidend für Fortschritte auf Gebieten wie der Stammzellforschung, der Gewebezüchtung, der Immunologie, oder der Krebstherapie, um nur einige Beispiele zu nennen.

In diesem Kontext ergeben sich zahlreiche bislang ungeklärte Fragen, die Herausforderungen auch an die mathematische Modellierung darstellen. So stellt sich zum Einen die Frage, welche Faktoren dazu führen, dass eine Zelle einen bestimmten Zelltyp annimmt, und wie sich diese kontrollieren lassen; zum Anderen, wie sich unterschiedliche Zelltypen in einer Zellpopulation bestimmen lassen. Die zelltypbestimmenden Faktoren finden sich vor allem auf der Ebene der Genregulation, welche in jeder individuellen Zelle abläuft. Äußere Einflüsse, wie biochemische Stimuli, beeinflussen die Genregulation, die wiederum für den Zelltyp ausschlaggebend ist. Die Bestimmung des Zelltyps findet dann meist aufgrund zelltypspezifischer Merkmale, etwa sogenannter Oberflächenmarker – Genexpressionsprodukte, die sich auf der Zelle nachweisen lassen – statt. Auf Zellpopulationsebene hingegen verändert sich die Anzahl von Zellen eines bestimmten Zelltyps laufend, da Zellen sich sowohl teilen als auch absterben. Um diese Proliferationsprozesse zu analysieren, werden in Experimenten die Zellen mit speziellen Labelingstoffen versehen, deren Konzentrationsänderung wiederum Rückschlüsse auf die Proliferationsraten zulässt.

Um Genregulationsprozesse zu verstehen, werden häufig dynamische differentialgleichungsbasierte Modelle verwendet. Jedoch gibt es bei der Wahl eines geeigneten Modellierungsansatzes einen Konflikt zwischen niedriger Komplexität, welche eine Vielzahl analytischer Methoden zulässt, jedoch nicht detailliert genug sein kann, und hoher Komplexität, welche eine realistischere Nachbildung ermöglicht, sich jedoch analytischen Methoden entzieht. Für die Modellierung von Zellpopulationen sind inzwischen diverse Modelle verfügbar, die dafür geeignet sind, aus den gängigen Proliferationsexperimenten mit großer Genauigkeit und Zuverlässigkeit die relevanten Parameter zu gewinnen. Jedoch gibt es bislang keine geeigneten Modelle für Populationen mit mehreren Zelltypen. Eine solche Modellklasse sollte sowohl die Zelltypen als auch die Prozesse der Zellteilung und des Labelingverfahrens berücksichtigen.

Die vorliegende Arbeit leistet wichtige Beiträge, um die angesprochenen offenen Probleme in der mathematischen Modellierung von Zelltypen lösen zu können:

- Die systematische Modellierung und Analyse eines Genregulationssystems wird demonstriert, und eine Konstruktionsmethode wird entwickelt, welche es ermöglicht, derartig erhaltene Ergebnisse auf ein detailliertes Genregulationssystem zu übertragen.

- Eine verallgemeinerte Modellklasse wird eingeführt, die sowohl Zelltypen und Zellteilungszahlen als auch die Labeleigenschaften berücksichtigt, und davon ausgehend werden spezifische Modellklassen für zwei gängige Labelingverfahren entwickelt.

Zu den behandelten Themenkomplexen wird jeweils anhand von Beispielen gezeigt, wie die vorgestellten Modellierungsmethoden in der Anwendung neue Erkenntnisse ermöglichen.

Einführung in das Thema

Genregulationsnetzwerke und genetische Schalter

Die Gene, die in jeder biologischen Zelle enthalten sind, entsprechen einer Art Bibliothek von Bauplänen. Wird ein bestimmtes Gen aktiviert, so wird der darin kodierte Bauplan ausgeführt und entsprechende Genexpressionsprodukte (mRNA, Proteine) werden produziert. Diese Genexpressionsprodukte wiederum bestimmen, wie sich die Zelle verhält und welche Eigenschaften sie ausbildet. Desweiteren können Genexpressionsprodukte die Expression weiterer Gene beeinflussen, indem sie diese zum Beispiel aktivieren oder auch unterdrücken. Dadurch entsteht ein komplexes Netzwerk von Interaktionen zwischen den in einer Zelle enthaltenen Genen. Einen Genexpressionszustand, der sich unter gleichbleibenden Bedingungen nicht mehr verändert, nennt man einen stabilen Zustand des Genregulationsnetzwerks.

Als genetischen Schalter bezeichnet man ein Genregulationsnetzwerk, welches für ein und dieselben Bedingungen mehrere stabile Zustände einnehmen kann. Das Konzept der genetischen Schalter dient als Erklärung dafür, warum und unter welchen Gegebenheiten die Genexpression zwischen getrennten stabilen Zuständen umschalten kann. Da unterschiedliche Zelltypen durch unterschiedliche Genexpressionszustände gekennzeichnet sind, und Zellen gleichen Typs durch das gleiche Genexpressionsmuster charakterisiert sind, entspricht solch ein Umschalten gleichzeitig einer Änderung des Zelltyps. Die Genexpression bestimmt also den Zelltyp.

Zellpopulationen und Labeling-Verfahren

Aufgrund der biologischen und experimentellen Gegebenheiten werden Zellen in der Regel in Zellpopulationen kultiviert. Eine Zellpopulation umfasst eine Vielzahl von Zellen, die nur selten alle dem gleichen Zelltyp angehören. Üblicherweise finden sich in einer Zellpopulation mehrere Zelltypen, die sich zum Beispiel darin unterscheiden, wie schnell sie sich vermehren oder sterben. Die Erfassung dieser Zelltypen geht häufig mit erheblichem experimentellen Aufwand einher oder ist gar unmöglich. Die Kenntnis darüber, wie viele Zellen welchen Zelltyps in der untersuchten Population vorliegen, und wie sich diese Zahlen während eines Experiments verändern, ist jedoch von entscheidendem Interesse etwa in der Stammzelltherapie, Krebsbekämpfung, und Immunologie.

Zur Quantifizierung der Zellproliferation, das heißt der Raten mit denen die Zellen sich vermehren und sterben, werden sogenannte Labeling-Experimente durchgeführt. Diese basieren darauf, dass der verwendete Label-Stoff (etwa ein spezieller Farbstoff, oder eine Markierung in der DNA) sich bei jeder Zellteilung auf die Tochterzellen aufteilt, und/oder bei der Zellteilung zusätzliches Label aus der Umgebung in die Tochterzellen aufgenommen wird. Durch Messung der Labelintensität der Zellen einer Zellpopulation erhält man Daten, die Rückschlüsse darauf zulassen wie oft sich eine typische Zelle in der beobachteten Zeit geteilt hat.

Mathematische Modellierung und Analyse von Zelltypübergängen

Die Systembiologie benutzt mathematische Modelle zur quantitativen Beschreibung biologischer Systeme und Prozesse. Für dynamische Vorgänge sind dies in der Regel Differentialgleichungsmodelle, die zur Analyse und Simulation herangezogen werden. Hierbei lassen sich gewöhnliche, stochastische und partielle Differentialgleichungen unterscheiden. Gewöhnliche Differentialgleichungen beschreiben dynamische Prozesse über die Zeit, stochastische Differentialgleichungen können zusätzlich Zufallsprozesse in der Dynamik abbilden. Partielle Differentialgleichungen hingegen dienen zur Beschreibung von Dynamiken, die neben der Zeit auch noch über weitere davon unabhängige Dimensionen stattfinden, wie zum Beispiel räumliche Ausdehnung oder Stoffkonzentrationen.

Für die Modellierung der Genexpression der einzelnen Zelle werden meist gewöhnliche Differentialgleichungen benutzt, die das genregulatorische Netzwerk und die daraus resultierende deterministische Dynamik abbilden. Daneben werden stochastische Differentialgleichungen verwendet, um zu untersuchen, inwiefern sich die Genexpression verändert, wenn Zufallsprozesse eine Rolle spielen. Ein stochastisches Differentialgleichungssystem wird sehr wahrscheinlich für jeden Simulationslauf eine andere Lösung liefern, da die Zufallsprozesse sich jedes Mal unterscheiden. Dementsprechend betrachtet man bei stochastischen Differentialgleichungsmodellen ein ganzes Ensemble von Lösungen, die wiederum einer Population von einzelnen Zellen entsprechen. Häufig nimmt man an, dass das mittlere Verhalten einer solchen Zellpopulation mit stochastischer Dynamik wieder dem gleicht, welches man aus dem entsprechenden deterministischen Differentialgleichungsmodell erhält, und verzichtet auf die aufwändigere stochastische Modellierung. Diese Annahme trifft aber nicht notwendigerweise zu, so dass unter Umständen stochastische Prozesse zu qualitativ anderem Systemverhalten auf Populationsebene führen. Ein Beispiel ist die Differenzierung von Stammzellen zu Knochenzellen, die erfahrungsgemäß nicht für alle Zellen einer Population erfolgreich erzielt werden kann – dieser Effekt lässt sich nicht mit einem deterministischen (gewöhnlichen) Differential-gleichungsmodell, jedoch mit einem stochastischen Differentialgleichungsmodell nachbilden.

Modelle genetischer Schalter stellen eine spezielle Form von Genregulationsmodellen dar, die zur Untersuchung konzeptioneller dynamischer Eigenschaften herangezogen werden. Ihre Komplexität befindet sich in der Regel noch in einem Rahmen, der etwa analytische Verfahren aus der Systemtheorie erlaubt. Genregulationsmodelle detaillierterer Form hingegen werden dazu benutzt, in Verbindung mit experimentellen Daten die Parameter des Zellsystems zu schätzen, und damit quantitative Aussagen und Prognosen über das Systemverhalten zu ermöglichen. So können zum Beispiel verschiedene experimentelle Szenarien in der Simulation ausprobiert und verglichen werden und eine optimale Vorgehensweise gefunden werden, um etwa die Differenzierung zu einem gewünschten Zelltyp zu erzielen. Beide Modellierungsansätze, die der Genschalter und die der detaillierteren Genregulationsmodelle, stehen jedoch bislang mehr oder weniger isoliert für sich. Desweiteren ist wenig darüber bekannt, welche Genschalterdynamiken zu bestimmten Populationsdynamiken führen können.

Auch für die Modellierung auf Populationsebene existieren bislang hauptsächlich zwei getrennte Ansätze: Betrachtet man die Anzahl von Zellen unterschiedlicher Zelltypen, so werden Systeme gewöhnlicher Differentialgleichungen verwendet. Ist jedoch ein kontinuierlicher Wert von Interesse, wie etwa die Genexpression oder die Konzentration eines Labels, so führt dies auf Systeme partieller Differentialgleichungen. Deren numerische Lösung ist nur unter hohem Rechenaufwand möglich, was Simulationen ineffizient und damit Parameterschätzung praktisch unmöglich macht. In den vergangenen Jahren wurden Populationsmodelle entwickelt, die zwar von einem System partieller Differentialgleichungen ausgehen, aber die sich in eine analytisch lösbare partielle Differentialgleichung und ein System gewöhnlicher Differentialgleichungen zerlegen lassen. Diese Modellklasse ist von großer Bedeutung, da sie zum Einen gängige bisherige Modelle beinhaltet und damit deren Lösung vereinfacht, zum Anderen eine deutlich effizientere Simulation und Analyse ermöglicht. Um nun Zellpopulationen mit mehreren Zelltypen modellieren zu können, ist die Verbindung beider bisher existierenden Modellierungsansätze notwendig. Für eine solche Modellklasse, die eine erhöhte Komplexität mit sich bringt, fehlt bisher ein vereinheitlichter Ansatz sowie effiziente Lösungsmethoden.

Forschungsbeiträge und Gliederung der Arbeit

Diese Arbeit widmet sich der mathematischen Modellierung von Zelltypübergängen. Hierfür werden sowohl die Genregulationsebene als auch die Zellpopulationsebene betrachtet und es wird gezeigt, wie beide Modellierungsebenen an der Schnittstelle zusammengeführt werden können. Im Folgenden wird eine Übersicht über die vorliegende Arbeit gegeben, und die zentralen Forschungsbeiträge darin erläutert.

Kapitel 1: Einleitung Im ersten Kapitel wird einleitend motiviert, weshalb die mathematische Modellierung von Zelltypübergängen fundamental für das quantitative Verständnis wichtiger biologischer Prozesse ist. Es wird kurz dargestellt, welche Ansätze und Beiträge es bisher gibt, und weshalb die Entwicklung neuer modellbasierter Methoden weiterhin notwendig ist. Es wird die Hypothese formuliert, dass ein ganzheitlicherer Modellierungsansatz zum besseren quantitativen Verständnis von Zelltypübergängen beitragen kann. Die Forschungsbeiträge, welche die Arbeit in den folgenden Kapiteln liefert, werden kurz erläutert und in den Zusammenhang der übergeordneten Fragestellung gesetzt.

Kapitel 2: Modellierung und Analyse eines Genschalters in der Zelldifferenzierung Im zweiten Kapitel wird exemplarisch ein generisches Genschaltermodell für einen Zelltypübergang in der Stammzelldifferenzierung entwickelt und analysiert. Es wird gezeigt, wie ein solches Modell in Kombination mit systemtheoretischen und simulationsbasierten Methoden dazu beitragen kann, genregulatorische Faktoren für Zelltypübergänge zu verstehen. Ein geeignetes Modell wird nach systemtheoretischen Kriterien entwickelt, und daraufhin mittels Multistabiliäts- und Bifurkationsanalyse sowie Simulationen untersucht. Dies liefert zum Beispiel Erkenntnisse darüber, welche experimentellen Bedingungen zu erwünschtem beziehungsweise unerwünschtem Systemverhalten führen können. Die Erweiterung zu einem stochastischen Modell ermöglicht es, Effekte zu untersuchen, die erst auf der Ebene der Zellpopulation auftreten. Die Modellentwicklung und Analyse sind möglichst allgemein gehalten, so dass sowohl die Vorgehensweise als auch die erhaltene Modellstruktur einen Entwurf für ähnliche Problemstellungen bieten.

Kapitel 3: Konstruktion von multistabilitätsäquivalenten Genregulationsmodellen Das dritte Kapitel beschäftigt sich mit der Frage, wie aus einem generischen Genschaltermodell (wie im vorherigen Kapitel vorgestellt) ein detaillierteres Genregulationsmodell konstruiert werden kann, und wie die aus dem Genschaltermodell gewonnenen Resultate auf ein solches detailliertes Modell übertragen werden können. Hierzu wird in dieser Arbeit das allgemeine Konzept der Multistabilitätsäquivalenz zwischen Genregulationsmodellen unterschiedlicher Dimension eingeführt, und eine Konstruktionsmethode für eine breite Systemklasse vorgestellt, mit der aus niedrigdimensionalen Modellen multistabilitätsäquivalente höherdimensionale Modelle erstellt werden können. Die Konstruktionsmethode wird exemplarisch dafür verwendet, aus einem Genschaltermodell ein detailliertes Modell für ein Genregulationsnetzwerk zu entwerfen, welches in der Stammzelldifferenzierung eine wichtige Rolle spielt und an experimentelle Daten angepasst werden kann.

Kapitel 4: Modellierung von proliferierenden Multizelltyp-Populationen in Labeling-Experimenten Dieses Kapitel präsentiert eine allgemeine Modellklasse für proliferierende Multizelltyp-Populationen in Labeling-Experimenten. Diese neue Modellklasse berücksichtigt sowohl verschiedene Zelltypen, als auch die Teilungszahl der Zellen sowie die Konzentration des Labels, und verallgemeinert damit zahlreiche existierende Modelle. Zugleich stellt diese Modellklasse wichtige Informationen bereit, die etwa benötigt werden, um Modellprognosen mit experimentellen Daten abzugleichen. Es werden Lösungsansätze für gängige Modelle aus dieser Modellklasse entwickelt und der Zusammenhang mit bereits existierenden Populationsmodellen diskutiert. Weiter wird untersucht, wie die im zweiten und dritten Kapitel behandelten Genschalterdynamiken in zelltypstrukturierte Populationsmodelle eingehen können.

Kapitel 5: Modellierung von CFSE-markierten Multizelltyp-Populationen Ausgehend von der im vorherigen Kapitel entwickelten allgemeinen Modellklasse wird ein Modell für das Labelingverfahren mit Carboxyfluoresceinsuccinimidylester (CFSE) spezifiziert. Die labelspezifischen Eigenschaften werden ausgenutzt, um die Modellgleichungen zu konkretisieren, diese in einfachere Teile zu zerlegen, und eine effiziente Analyse und Simulation zu ermöglichen. Dieses Modell für CFSE-gelabelte Multizelltyp-Populationen erlaubt es, zelltypspezifische Proliferationseigenschaften zu untersuchen, was am Beispiel einer differenzierenden Stammzellpopulation demonstriert wird.

Kapitel 6: Modellierung von BrdU-markierten Multizelltyp-Populationen Das sechste Kapitel stellt ein spezifisches Modell für das Labelingverfahren mit Bromdesoxyuridin (BrdU) vor, das ebenfalls sowohl verschiedene Zelltypen als auch die volle Labelverteilung in der Zellpopulation abbilden kann. Erneut ausgehend von der zuvor eingeführten allgemeinen Modellklasse werden nun die Eigenschaften des BrdU-Labelingverfahrens berücksichtigt, um wiederum ein spezifisches Modell mit effizienter Analyse und Simulation zu gewinnen. Für die Aufnahme des Labels BrdU und dessen Weitergabe an Tochterzellen während der Zellteilung wird ein mathematisch rigoroses Modell entwickelt, das die probabilistischen Eigenschaften der Chromosomenverteilung abbildet. Durch die explizite Modellierung der Labelkonzentration kann dieses Modell für BrdU-markierte Multizelltyp-Populationen die volle Information aus Daten verwenden. So können mit diesem neuen Modell präzisere Aussagen über Proliferationseigenschaften von untersuchten Populationen getroffen werden, wie anhand beispielhafter Untersuchungen typischer Daten von proliferierenden Zellpopulationen gezeigt wird.

Kapitel 7: Fazit Das letzte Kapitel schließt mit einer Zusammenfassung und einem Ausblick. Die in dieser Arbeit erhaltenen Ergebnisse werden nochmals in Bezug zur übergeordneten Fragestellung, der mathematischen Modellierung von Zelltypübergängen, gesetzt. Davon ausgehend werden offene Fragen und potentielle weitere Entwicklungen diskutiert.

Chapter 1

Introduction

1.1 Research motivation

For many decades now, mathematical modeling has enabled the quantitative understanding of complex systems in numerous disciplines. One of these disciplines is biology, which is dedicated to the study of living organisms and as this is confronted with inconceivably complex processes. The questions that biology is pursuing nowadays range from a mechanistic understanding of how a complete organism emerges, to predicting which diseases may be genetically predetermined. Such questions can hardly be answered by purely experimental and empirical approaches any more. To tackle these challenging problems, and simultaneously resolve knowledge at ever higher detail, biology has greatly benefit from mathematical and model-based approaches for several decades now since the hallmark works of Hodgkin and Huxley (1952); Turing (1952); von Foerster (1959); Welch (1947). The emerging field has been termed as systems biology or mathematical biology, devoted to the quantitative understanding of biological processes via mathematical models (Klipp et al., 2000).

As sophisticated mathematical models have become indispensable, their development requires to exploit advanced methods and approaches from various fields such as system dynamics, systems and control theory, statistical analysis, and simulation technology, without being complete. The successful development of mathematical modeling frameworks from a methodological viewpoint needs to balance between generality, which makes modeling approaches applicable to a broad range of problem settings, and specificity. Specifications in this context may arise from two different sides, according to the interdisciplinarity of this field: Biology, on the one hand, provides certain requirements what models "should be able to", and assumptions that "make sense". Mathematical modeling, on the other hand, needs restrictions to certain system classes, in order to handle and solve the posed problems.

This thesis focuses on the mathematical modeling and analysis of cell type transitions, which are fundamental to all living organisms but still far from being fully understood. Cell type transitions occur whenever a cell, or many cells within a cell population, change their cell type. Popular examples are: the differentiation of nonspecialized stem cells into more specialized cell types, the transformation of healthy cells into pathological cells due to disease, or vice versa of harmful cells into a harmless cell type upon medical treatment. The quantitative understanding of cell type transitions is crucial for advancing highly relevant fields such as stem cell research, tissue engineering, immunology, and cancer therapies, to name just a few. A multitude of questions arises in this context which provide mathematical modeling problems to be addressed in a general and yet suitably specific way.

1.2 Overview on the research topic

1.2.1 Cell type transitions

Each organism is constituted by multiple different cell types – the human body, for example, comprises about 200 cell types (Alberts et al., 2000). Each cell type in turn is characterized by a specific gene expression pattern, although the cells within the same organism usually carry identical genes, that is the same Deoxyribonucleic acid (DNA). From time to time, cells undergo a transition to a different cell type, for example if stem cells become more specialized mature cells. These cell type transitions are driven by a change in the cell type-determining gene expression state. To investigate these determinants of cell types, studies of gene expression measure the concentration of gene expression products such as messenger Ribonucleic acid (mRNA), proteins or transcription factors.

Cell types then determine the composition of populations such as tissue, organs, and whole organisms. But even in smaller and more defined cell populations such as in the Petri dish, often several cell types are present. Studied cell populations therefore are in most cases composed of several subpopulations of which each has its own cell type-specific properties, but which can usually not be distinguished by the applied experimental approaches. One of the most fundamental questions about a cell population concerns its proliferation, that is the rates at which cells divide and die. For these cell population studies, commonly labeling techniques are employed which are based on a biochemical marker such as Carboxyfluorescein succinimidyl ester (CFSE) (Hawkins et al., 2007; Lyons and Parish, 1994; Lyons, 2000) or Bromodeoxyuridine (BrdU) (Gratzner, 1982). Importantly, these labeling techniques do not imply to distinguish between distinct cell types, but this relevant information has to be revealed from the data by additional approaches.

1.2.2 Established modeling approaches

For quantitatively describing and modeling cell type transitions, mainly two different approaches can be distinguished: To understand how the cell type of individual cells is determined by gene expression, gene regulatory network models have been established, with the work of Ferrell and Machleder (1998) representing one popular and early example. To monitor and control the dynamics of cell populations, population models are utilized since they have been introduced for example by von Foerster (1959).

Gene regulatory network models are motivated from different intentions: On the one hand, conceptual or core motif models are low-dimensional, abstract representations which mainly serve to reproduce and explain generic properties of cell type transitions, such as underlying gene regulatory dynamics that are responsible for switch-like behavior. For example, nonlinear ordinary differential equation models are commonly employed for this purpose, as by Huang et al. (2007); Narula et al. (2010); Roeder and Glauche (2006); Xiong and Ferrell (2003), to name just a few. On the other hand, data-driven gene regulatory network models usually comprise larger sets of genes to capture the details of realistic gene interaction networks. Since considering a larger number of genes means that the model has to be of higher dimension, the dynamics are usually restricted to be Boolean or linear in order to retain a tractable complexity, see examples by Krumsiek et al. (2011); Weber et al. (2013a).

Cell population models in turn are also developed and applied with different purposes: Some cell population models aim to capture the proliferation properties of the population under study in detail, and for this account for cell types with distinct proliferation rates (Bonhoeffer et al., 2000; De Boer and Perelson, 1995; De Boer and Noest, 1998; De Boer et al., 2003a; Ganusov and De Boer, 2013; Glauche et al., 2009; Jones and Perelson, 2005; Parretta et al., 2008; Wilson et al., 2008). Others emphasize the relationship to data from labeling experiments, and allow to reconstruct the full measurement information (Banks et al., 2010; Hasenauer et al., 2012b,a; Luzyanina et al., 2007b, 2009; Schittler et al., 2011; Thompson, 2011). Although both aspects are without doubt highly relevant, there is still a lack of models that consider both, the cell type structure as well as the full label measurements.

1.2.3 Open problems

In view of the existing approaches for modeling and analyzing cell type transitions, there are several open problems as pointed out in the following.

Regarding the development of gene regulatory network models, core motif models on the one hand and detailed models on the other hand are at present rather isolated from each other. While systems-theoretic analysis can usually operate on low-dimensional core motif models, matching to experimental data requires more realistic and hence higher-dimensional models. In this way, results obtained from core motif models can not be transferred directly and rigorously to more realistic gene regulatory network models of higher dimension.

Concerning cell population models, there is lack of models that account for cell types and cell type-specific parameters, and at the same time allow to exploit the full information from data. Making the whole measurement information available, however, is especially important when cell type subpopulations should be revealed.

For common labeling techniques, specific models for labeled proliferating cell populations are available. However, for CFSE labeling no model exists that accounts for cell types, and for BrdU none exists that accounts for the full label information. Moreover, for BrdU labeling and similar techniques that apply DNA labels, no rigorous model for the segregation of DNA label is known so far. Thus, it can not be reproduced correctly how cells inherit the label, which is crucial for deducing the proliferation properties of interest from the label measurements.

This thesis hypothesizes that different modeling approaches – core motif models, detailed gene regulatory models, population models with cell type-structure and label resolution – are essential to promote the understanding of cell type transitions. A more holistic modeling framework, which accounts for these relevant aspects, is expected to allow for new quantitative insights into how cell type transitions are determined and can be monitored. In this thesis, we aim to contribute a step towards a more comprehensive and holistic mathematical modeling framework for cell type transitions.

1.3   Contribution of this thesis

In this thesis, we develop novel mathematical modeling approaches that are dedicated to the challenges pointed out beforehand. The presented framework supports the development of suitable models, their mathematical analysis, and their numerical simulation. The modeling methods thereby address problems arising at distinct relevant levels – modeling the gene expression (which characterizes cell types and cell type transitions) and modeling the cell population (where cell types emerge as subpopulations). We provide a brief overview of the major contributions of this thesis.

As has been argued above, the two modeling approaches of low-dimensional core motif models and high-dimensional gene regulatory network models are rather isolated.

Contribution 1 *In this thesis, we contribute to solving this problem for a broad system class via a new **construction method for dynamical gene regulatory network models**: Given a low-dimensional dynamical model and the qualitative structure of a high-dimensional gene regulatory network, the construction methods yields a high-dimensional dynamical model that exhibits the same multistability properties, while also complying the postulated interaction structure. This method allows to use core motif models as blueprints for more realistic high-dimensional models of gene regulation.*

The gene regulatory dynamics that determine cell types can be incorporated for example into a population balance model, to obtain a model for arising cell type dynamics on population level. Common experiments to study cell population dynamics rely on labeling techniques but usually do not explicitly resolve cell types, and existing models do only account for either cell types or the full label distribution.

Contribution 2 *We propose a **generalized model class for proliferating populations that accounts for multiple cell types and the full label distribution**. This model class contains many established models as special cases. The unifying approach allows to develop analysis and solution tools on a general level, independent of specific applications. For example, we present analytical solutions for several cases, and derive bounds for the truncation error if not the full model can be simulated. Furthermore, it offers a link to the population balance models that in turn can also consider the dynamics of gene regulation. Providing the full label distribution is an important contribution since this allows to exploit the full information from data, and thus facilitates more reliable and exact parameter estimates for proliferating multi-cell types populations. The presented model class is, to the best of the author's knowledge, the first one that unifies all three aspects of cell type dynamics, proliferation dynamics, and label dynamics.*

In order to unravel cell type dynamics from data of labeling experiments, one needs sophisticated models that are tailored to the respective labeling technique. At the same time, a model needs to allow for an efficient solution and simulation.

Contribution 3 *We present two novel **cell population models specifically tailored to the labeling techniques of BrdU and CFSE**, respectively, which are specifications of the general model class. Both specified models can be solved efficiently due to a decomposition approach, as we prove in the respective chapters. With this, the advantages and tools of the presented* *model class are readily available for these common labeling techniques. These models can be used for modeling and estimating parameters of BrdU- and CFSE-labeled multi-cell type populations, as is also demonstrated in examples.*

Contribution 4 *Advancing the modeling of populations in DNA labeling experiments, we present a* **mathematically rigorous model for the segregation of labeled and unlabeled chromosomes** *from mother to daughter cells. This model is, to the best of the author's knowledge, the first mathematically rigorous one for the relationship between a cell's label content and its number of undergone divisions for DNA labeling experiments. The model of chromosome segregation is incorporated into the specific cell population model for BrdU labeling. This allows to derive a more realistic model for the label distribution in proliferating cell populations, and thus also a more exact and more reliable estimation of the proliferation properties especially when multiple cell types are present.*

The presented modeling and analysis methods contribute to a more holistic framework for the quantitative understanding of cell type transitions. While supporting and exploiting systems-theoretic and mathematical analysis approaches, the proposed modeling framework at the same time regards and even profits from specific properties of the considered biological systems.

1.4 Outline of this thesis

In **Chapter 2**, a model for a gene switch that determines a cell differentiation process is developed and analyzed. This exemplifies how core motif models of gene regulation can elucidate generic mechanisms and dynamical properties of cell type transitions, whereas the presented approach may be viewed as a blueprint for modeling and analyzing cell type-determining gene switches.

To provide a link from core motif models to more realistic, detailed gene regulatory network models, we present a construction method for multistability-equivalent gene regulatory network models of different dimensionality in **Chapter 3**. The construction procedure allows to derive a high-dimensional dynamical model which exhibits the same multistability properties as a given low-dimensional model, and thereby allows to transfer results from generic to more detailed models of gene regulatory networks.

After these contributions on modeling the gene regulation of cell type transitions, we turn toward modeling cell populations which comprise multiple cell types, and where cell type transitions are viewed on the population level. For this purpose, in **Chapter 4** we introduce a general model class for proliferating multi-cell type populations which are commonly studied by labeling experiments. The presented model class is as general as possible, such that methodological modeling approaches can be developed at this level, independent of the specific application and experimental setting.

Specific experimental settings are then considered in the following two chapters, to derive models tailored to common experimental labeling techniques. Based on the general model class introduced beforehand, **Chapter 5** presents a model for CFSE-labeled multi-cell type populations. The model accounts for cell types, division numbers, and the label concentration at the same time. Its applicability is demonstrated by studying a population of proliferating and differentiating stem cells.

In **Chapter 6**, a model for BrdU-labeled multi-cell type populations is developed. Again, this model is derived from the previously introduced general model class. As the label BrdU is carried in the DNA of cells, the comprehensive model development is preceded by rigorously modeling and analyzing the process of DNA label segregation. Combining the effects of label uptake and inheritance, we finally derive a cell population model that accounts for cell types, division numbers, and the measured label intensity at the same time. This allows to exploit the full information from BrdU data and promises new insights into labeling properties as well as more accurate estimates of proliferation properties, as we demonstrate in model-based studies on BrdU data.

Chapter 7 concludes with a summary of the key results of the thesis, and gives an outlook on open questions and possible future work.

Chapter 2

Modeling and analysis of a gene switch determining cell differentiation

In this chapter, we present the development and analysis of a model for a gene regulatory network that determines a cell type transition within the differentiation of mesenchymal stem cells. The model developed here can be viewed as an exemplary representative of typical problems in cell differentiation, and the according methods and tools are of general validity to similar problems. In Section 2.1, we provide an overview on the general background of cell differentiation and mesenchymal stem cells in particular, and briefly review available models for cell differentiation, after which we formulate the problem to be addressed in the following. In Section 2.2, we develop the model of a gene switch determining the cell type transition from a progenitor cell to one of two possible committed cell types. We perform stability and bifurcation analyses that reveil generic properties of the system, and simulations of distinct differentiation scenarios which serve to demonstrate the effects of stimulus inputs on the cell system. Section 2.3 presents an extension to a stochastic model, which is exploited to derive gene switch dynamics on the cell population level. The chapter concludes with a summary and discussion in Section 2.4.

This chapter is based on Schittler et al. (2010).

2.1 Background and problem formulation

2.1.1 Biological background of cell differentiation

Cell differentiation is indispensable to life, since it is fundamental for the emergence of a whole organism, composed of multiple specialized cell types, from a single fertilized egg cell. The process of cell differentiation can be described as a sequence of cell type transitions: Initially the cell is a multipotent progenitor or stem cell, meaning that its cell fate is not yet determined, but there exist several possibilities (Alberts et al., 2000; Ralston and Rossant, 2005). During differentiation the cell becomes more and more specialized by traversing several cell type transitions, for example from a multipotent cell, through progenitor and subsequently more committed cell types, until it finally ends up as a fully differentiated cell. The stability of the cell type which is adopted eventually is thereby of high importance for the maintenance of the organism.

The question of what triggers cell differentiation, what determines and eventually ensures the stable maintenance of a cell type has been a major topic in developmental biology for many decades. Waddington's epigenetic landscape (Waddington, 1956, 1957) hallmarks the onset of viewing cell differentiation as a dynamical system, and this view has been pursued until today (Bar-Yam et al., 2009), opening up the whole variety of methods from systems dynamics and mathematical systems theory. According to this view, cell differentiation is described by a trajectory through a landcape reflecting the energy potential of the system, as imposed by the dynamical gene regulatory system. The concept of Waddington's epigenetic landscape has been discussed in consideration of current experimental findings, for example by Balázsi et al. (2011); Glauche (2010); Goldberg et al. (2007), or with special focus on the relationship to systems dynamics and systems theory by Ferrell (2012).

With the flourishing of stem cell research in recent years, there has also emerged wide interest in understanding cell differentiation from a systems theoretic point of few. Due to their multipotency, stem cells promise to offer a high potential for medical therapies and tissue engineering. A particular stem cell type are mesenchymal stem cells which will serve as the example under study here. Mesenchymal stem cells (MSCs) can be derived from adults and can give rise to various cell types included for example in bone and cartilage (Baksh et al., 2004; Barry and Murphy, 2004; Frith and Genever, 2008; Heino and Hentunen, 2008). Therefore, special interest concerns their targeted application for bone or cartilage transplants. To make MSCs available for such tissue engineering applications, it is crucial to control the eventual commitment towards either a bone (osteogenic) or cartilage (chondrogenic) cell type. However, scientists are far from a detailed knowledge of the processes governing cell fate commitment, differentiation and maintenance of completely differentiated cell types.

The challenges in stem cell research are to identify stimuli or cocktails of stimuli that trigger the respective genes in order to end up with a desired cell type. Therefore, the gene regulatory network has to be revealed, effects of stimuli have to be evaluated, and alternative experimental protocols have to be compared with respect to their differentiation success. Systems biological approaches in combination with mathematical and systems theoretical modeling and analysis tools may offer a great potential in promoting the quantitative understanding of cell differentiation.

2.1.2 Available models for cell differentiation

Following and extending Waddinton's concept (Waddington, 1956, 1957) and supported by experimental findings, cell differentiation is viewed as a process of subsequent type transitions (Alberts et al., 2000): A cell type is characterized by a type-specific gene expression pattern, and as such can be interpreted as a stable steady state in the space of possible gene expression levels (Huang, 2009). Then, at each transition step, gene regulatory dynamics determine which of the possible cell types is adopted. Such a view has been formulated in mathematical modeling frameworks for example in terms of binary switch modules (Foster et al., 2009).

The demand for a more quantitative understanding of cell differentiation and its gene regulation has created vital research in mathematical modeling of cell differentiation. Depending on the problem at hand, for example ordinary differential equations (ODEs) or stochastic models provide an excellent approach to represent and analyze the dynamics of gene regulation in stem cell fate (Peltier and Schaffer, 2010) and are commonly used: During the recent decade, systems biology and dynamic models of cell differentiation have greatly improved understanding of certain types of stem cells such as embryonic stem cells (Glauche et al., 2010), hematopoietic stem cells (Chickarmane et al., 2009; Huang et al., 2007; Laslo et al., 2006; Narula et al., 2010; Roeder and Glauche, 2006; Palani and Sarkar, 2009), pancreatic cells (Zhou et al., 2011), preadipocytes (Coskun et al., 2010), T cells (Hong et al., 2012) or B cells (Martinez et al., 2012).

For other cell types, however, the key determinants guiding differentiation are still poorly known. Regarding MSCs, for example, some models have been developed for the differentiation into one specific lineage, and assuming linear gene regulation dynamics (Weber et al., 2013a,b). Only few studies are dedicated to modeling the gene regulatory dynamics of several competing MSC lineages (Assar et al., 2012; MacArthur et al., 2008). These models of gene regulation all struggle with the complexity and simultaneously sparse knowledge of the involved processes, contrary to the modeling of metabolic networks which can often benefit from more quantitative knowledge of biochemical reactions. Thus, dynamical models of gene regulation usually simplify and reduce the considered genes to a low number of key regulators and their functional relationships (Cinquin and Demongeot, 2005; Huang et al., 2007; Roeder and Glauche, 2006; Palani and Sarkar, 2009; Zhou et al., 2011). To extend the insight that can be obtained from such a generic model, the next Chapter 3 provides a method to expand such a model to a more detailed gene regulatory network model, which can incorporate a larger set of genes known to be involved in the process.

2.1.3 Modeling approaches for gene expression dynamics

As mathematical models for gene expression dynamics have been widely established, we provide a brief overview of important model classes.

A common way to model gene expression dynamics is via ordinary differential equation (ODE) models (Klipp et al., 2000; MacArthur et al., 2009). These models describe the concentration change of gene expression products in a single cell under deterministic dynamics,

$$\dot{x} = f(x, u), \tag{2.1}$$

with nonnegative state variables $x \in \mathbb{R}_+^n$, nonnegative inputs $u \in \mathbb{R}_+^m$, and the dynamics given by the vector field $f : \mathbb{R}_+^n \times \mathbb{R}_+^m \to \mathbb{R}_+^n$. The state variables x commonly represent the expression levels of the according genes, whereas the inputs u correspond to biochemical stimuli that influence gene expression. As such models are given as a system of ODEs (2.1), they can be solved via suitable numerical solvers. Besides solving such an ODE model, the study of its steady states is of the same interest. The stable steady states define the cell type-specific gene expression and in this way provide a partitioning of the state space into regions corresponding to distinct cell types.

In order to incorporate also stochastic dynamics, stochastic differential equation (SDE) models are employed, leading to the Chemical Langevin equation (CLE) (Higham, 2008; Wilkinson, 2006). These models describe gene expression dynamics, again in a single cell, from both deterministic and stochastic processes,

$$dX_t = f(X_t, U_t)dt + g(X_t, U_t)dW_t, \tag{2.2}$$

with nonnegative state variables $X_t \in \mathbb{R}_+^n$, nonnegative inputs $U_t \in \mathbb{R}_+^m$, the drift term $f : \mathbb{R}_+^n \times \mathbb{R}_+^m \to \mathbb{R}_+^n$, the diffusion term $g : \mathbb{R}_+^n \times \mathbb{R}_+^m \to \mathbb{R}_+^n$, and W_t a Wiener process.

Numerous methods exist for the numerical solution of SDEs and are discussed, for example, by Gillespie (1977, 2001); Higham (2001, 2008); Kloeden and Platen (1999). As the classical notion of stability does not hold any more under stochastic dynamics, the concept of stability in probability provides a complementary approach (Chas'minskij, 1980; Ignatyev and Mandrekar, 2010). Solving an SDE model numerically many times leads to an ensemble of stochastic trajectories, representing an ensemble of cells (Stamatakis, 2010) under stochastic dynamics.

In contrast to single cell models and ensemble models of single cells, population balance equation (PBE) models directly describe the dynamics on cell population levels (von Foerster, 1959; Fredrickson et al., 1967; Mantzaris, 2007; Ramkrishna, 2000; Stamatakis, 2010; Tsuchiya et al., 1966) which consist of partial differential equation (PDE) or even integro-partial differential equations. Common PBE models consider either deterministic gene expression dynamics and cell division (Mantzaris, 2007), or deterministic and stochastic gene expression dynamics (Ramkrishna. 2000), but usually not all three aspects together.

2.1.4 Problem formulation

In this chapter, we aim to derive a model for a cell type transition step within the differentiation of mesenchymal stem cells. More precisely, the considered cell type transition concerns a stage where the cell is already specialized as a progenitor for osteogenic and chondrogenic cells, but the eventual commitment towards either the osteogenic or the chondrogenic cell type is impending.

With the development of this model, we aim to approach the following challenges arising in modeling of cell differentiation:

- The detailed biochemical processes are not known, but a mathematical model of the gene regulatory dynamics is desired.

- Although the obtained model will be generic instead of giving quantitatively exact predictions, it should provide explanations for the major characteristics of cell differentiation.

- Additional cell differentiation effects may arise under stochastic dynamics, and to investigate these, a suitable model is needed.

The first problem addressed in this chapter concerns the deterministic dynamics and can be formulated, generally speaking, as the following problem:

Problem 2.1 *(Derivation and analysis of a deterministic gene switch model for cell type transitions of stem cells) Construct a gene switch model, (2.1), that can reproduce the cell types (in terms of steady states) and cell type transitions observed in a specific stage of cell differentiation.*

The second problem is dedicated to incorporating also stochastic dynamics, and to derive models for uncovering dynamical effects arising at the cell population level.

Problem 2.2 *(Derivation of a stochastic model for cell type transitions of stem cells) Derive a stochastic gene switch model to study cell type transitions, in a specific stage of cell differentiation, on the population level.*

The chosen modeling approach will follow the formalism of ordinary differential equations (ODEs) for genetic switches in cell differentiation, and enlarge to stochastic differential equations (SDEs), both as introduced above. However, already the deterministic model contributes some new insight into cell differentiation.

The presented work provides important contributions in two different aspects:

- The conducted approach represents a comprehensive *modeling scheme* for the systematic model development and analysis which is especially suited to model gene regulatory dynamics when there is sparse knowledge of detailed biochemical processes. The modeling scheme reaches from model specifications, structure selection, formulation of kinetics and parametrization to the analysis of steady states and transitions between states, complemented by a stochastic model extension.

- The derived model proposes a *novel and general gene switch structure*, in which the two-steps process of a cell type transition within differentiation is implemented via two switch parts: A preswitch determines whether the cell will differentiate, whereas a fate switch determines to which cell type it will differentiate. The obtained gene switch structure represents a general and robust design principle, which also provides a connection between subsequent cell type transition steps and thereby supports the view of cell differentiation as subsequent steps of cell type transitions[1].

The presented approach may be pursued analogously for other applications in cell fate determination, especially the procedure of model specification, development, analysis, and stochastic extensions. What is more, the model itself is generic in the sense that it represents a mechanism that might be quite general and commonly found in cell type transitions. We thus suggest that our procedure of model development and analysis, although exemplified for a specific system, may serve as a blueprint for gene switch modeling in cell type transitions.

2.2 A gene switch model for cell differentiation

The specific biological system under consideration regards one of several steps in the differentiation of mesenchymal stem cells (MSCs): Progenitor cells originating from MSCs have already undergone several steps of specialization towards osteogenic-chondrogenic commitment, while they still remain bipotent in the sense that they can further differentiate into either osteoblasts (bone cells) or chondrocytes (cartilage cells) (Zhou et al., 2006; Heino and Hentunen, 2008).

2.2.1 Model specification

For constructing a mathematical model of a particular system, it is recommendable to first collect the known system properties – for example, observed from biological experiments – and second, to translate them into mathematical formulations, which can then be imposed as requirements on the model to be developed.

---

[1]Interestingly, after this work was first published by Schittler et al. (2010), similar design principles have also been exploited for modeling cell differentiation of other cell types by Zhou et al. (2011).

The here investigated biological system of differentiating MSCs provides several working assumptions for developing a model, which we will subsume in this subsection. The features of the system will then be exploited to specify an ODE model of the form (2.1). We will clarify the meaning of such an ODE model equation and its contents throughout the following, accompanied by interpretations in the biological sense.

Cell types and type-specific gene expression There are three cell types to be captured by the model, and each one is characterized by a type-specific pattern of gene expression. The three cell types that we aim to capture are: a progenitor (P), an osteogenic (O), and a chondrogenic (C) cell type. They are mutually exclusive, and should be the only steady states of the model.

As long as there is no additional influence on the gene expression, the cell types stably maintain their state. Thus, we require the model to capture these cell types $j \in \{P, O, C\}$ as stable steady states of the model equations. To be precise, that is as solutions $x^{*,j} = (x_P^{*,j}, x_O^{*,j}, x_C^{*,j})^T$ when no inputs are applied of

$$f(x^{*,j}, 0) = 0, \tag{2.3}$$

with the Jacobian $(\partial f(x, u)/\partial x)|_{x=x^{*,j}}$ having eigenvalues λ_k^j with $\Re(\lambda_k^j) < 0 \; \forall k$. Therein, $f(x, u)$ is the right hand side of the model equations that will be derived later on. In this way, demanding the model to capture these and only these cell types is translated into requiring the existence of exactly three stable steady states. Each steady state $x^{*,j}, j \in \{P, O, C\}$, is enclosed by an attractor basin

$$\Omega_j := \{x_0| \lim_{t \to \infty} x(t, x_0) = x^{*,j}\}, \tag{2.4}$$

where $x(t, x_0)$ is the solution of the dynamics given by an ODE model (2.1) to be derived, with $x(0) = x_0$.

Each cell type is recognized by a high level of a characteristic gene expression product or transcriptional regulator (TR), denoted x_P, x_O, or x_C respectively, but low levels of the other TRs. For example, RUNX2 or OSX are established as osteogenic TRs (Ryoo et al., 2006), whereas SOX9 is known as a chondrogenic TR (Zhou et al., 2006). For the maintenance of the progenitor cell, TRs may be Nanog (MacArthur et al., 2008) or the cytokine Tweak which acts pro-proliferative and differentiation-inhibiting (Girgenrath et al., 2006; Winkles et al., 2007). This can be formulated generally, requiring each steady state $x^{*,j}$ to have the type-specific gene expression level for cell type j with:

$$\forall i \neq j : x_j^{*,j} > x_i^{*,j}. \tag{2.5}$$

The relationship of TR levels and cell types is schematically depicted in Figure 2.1, where the axes denote the TR levels x_P, x_O, x_C. Relating to experimental data, these would be (as a rough measure) mRNA levels or (as a more precise measure) transcription factor activities from reporter genes. Accordingly, the model will have a three-dimensional state space, $n = 3$ in (2.1).

The cell type of the system can now be formulated via the system state's coordinates in the three-dimensional state space, using the definition of attractor basins (2.4) and the cell type-characteristic gene expression levels (2.5). This taken together, the cell types that the model should reproduce are defined as

$$
\begin{aligned}
\text{P} &\Leftrightarrow x \in \Omega_\text{P} \text{ and } x_\text{P}^{*,\text{P}} > x_\text{O}^{*,\text{P}} \text{ and } x_\text{P}^{*,\text{P}} > x_\text{C}^{*,\text{P}} \\
\text{O} &\Leftrightarrow x \in \Omega_\text{O} \text{ and } x_\text{O}^{*,\text{O}} > x_\text{P}^{*,\text{O}} \text{ and } x_\text{O}^{*,\text{O}} > x_\text{C}^{*,\text{O}} \\
\text{C} &\Leftrightarrow x \in \Omega_\text{C} \text{ and } x_\text{C}^{*,\text{C}} > x_\text{P}^{*,\text{C}} \text{ and } x_\text{C}^{*,\text{C}} > x_\text{O}^{*,\text{C}}.
\end{aligned}
\tag{2.6}
$$

Cell type transitions In differentiation experiments, biochemical stimuli are applied and it is observed that there occur certain transitions between the three cell types (Baksh et al., 2004; Heino and Hentunen, 2008), while others do not. The model should reproduce this by according state transitions induced by certain inputs. In terms of systems theory, a transition between two cell types means that the system is pushed out of the attractor basin Ω_{j_1} of one stable steady state j_1, and into the attractor basin Ω_{j_2} of another stable steady state j_2. Let us denote a transition from cell type j_1 to cell type j_2, with $j_1, j_2 \in \{\text{P}, \text{O}, \text{C}\}$, by $(j_1 \to j_2)$.

Formulating the biological observations about transitions, there should be only two possible transitions of differentiation, namely from the progenitor to either the osteogenic or the chondrogenic cell type,

$$
\Delta^\text{diff} = \{(\text{P} \to \text{O}), (\text{P} \to \text{C})\}.
\tag{2.7}
$$

Other transitions that are not observed in the experimental settings considered here, are de-differentiation

$$
\Delta^\text{de} = \{(\text{O} \to \text{P}), (\text{C} \to \text{P})\},
\tag{2.8}
$$

and trans-differentiation

$$
\Delta^\text{tra} = \{(\text{O} \to \text{C}), (\text{C} \to \text{O})\}.
\tag{2.9}
$$

Although these transitions may occur in other experimental settings, these alternative settings are not the focus of the model to be developed here.

Biochemical stimuli The cell type transitions, as long as viewing gene expression and cell differentiation as a deterministic process, should thereby not occur by chance, but upon suitable stimulation via biochemical components. The effect of such biochemical stimuli will be represented in the model by input components of the input u. It is assumed that there are three distinct inputs, thus $p = 3$ in (2.1), which can be applied separately or in combination with each other:

- A differentiation stimulus, u_D, should be required to enable the system to escape from the progenitor cell type. Commonly in differentiation experiments, certain biochemicals have to be provided irrespective of the desired cell type to promote differentiation. In MSC differentiation, these may be for example ascorbic acid, or dexamethasone.

- An osteogenic stimulus, u_O, should support differentiation into the osteogenic cell type. However, it is assumed that the osteogenic stimulus alone is not sufficient to induce differentiation. For example, in MSC differentiation β-glycerophosphate is contained in osteogenic differentiation media, but is applied in combination with ascorbic acid and dexamethasone (Kawaguchi et al., 2005; Shui et al., 2003).

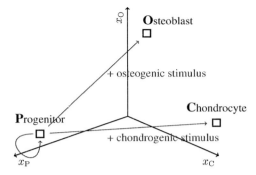

Figure 2.1: Cell types (□) and transitions between them (→) in the cell differentiation of osteo-chondro progenitors. Each cell type is recognized by a high level of one of the three TRs (three axes). The possible transitions from the progenitor to a differentiated state take place upon corresponding stimuli.

- A chondrogenic stimulus, u_C, should support differentiation in a similar way into the chondrogenic cell type, while again being not sufficient on its own to induce differentiation. For differentiating MSCs into chondrogenic cells, TGFβ is applied in combination with ascorbic acid and dexamethasone[2].

This requirement is in perfect agreement with the chosen model class of ODEs, as can be seen by comparing the biological interpretation to the systems theoretic interpretation which is provided in parenthesis: Given a certain cell state as a starting point (an initial condition of the ODE), the future evolution of the system dynamics are fully determined by this initial state and the dynamical properties (the ODE has exactly one solution). If, in addition, the starting point is a stable cell type (stable steady state), the cell will remain in this state (the stable steady state will not be left, by its definition, even for small perturbations, Slotine and Li (1991)). For the cell state to transit to another cell type (the trajectory of the system to converge to a different stable steady state), it is necessary that the currently maintained state does not exist as a stable cell type (stable steady state) for a sufficiently long time any more (namely, until the system trajectory has escaped the original attractor basin of this stable steady state). This is indeed the case for certain types of bifurcations, which will be analyzed later on and thereby serve to ensure the required properties.

Interactions between genes   Gene regulatory networks as they are found in cells are understood as networks of interacting genes, instead of just sets of isolated genes. To ensure a certain degree of interaction in the model, we claim that between any two genes considered in the model, there must be at least one interaction. This can be formulated by requiring the structure of the model to be weakly connected in a graph theoretical sense.

---

[2]experimental protocols, G. Vacun and J. Hansmann, Fraunhofer Institute for Interfacial Engineering and Biotechnology (IGB), Stuttgart

Let us briefly summarize the requirements on the model as outlined above:

(R1) The model should have three state variables, representing the expression level of three cell type specific TRs.

(R2) The model should have three stable steady states, each corresponding to one of three cell types, and they should be characterized by one TR expressed at high level, the other two at low levels.

(R3) The model should have three inputs that represent biochemical stimuli enhancing or suppressing the expression of a TR.

(R4) A differentiation stimulus of sufficiently high value and sufficiently long (finite) duration should be necessary to enable differentiation to any of the two differentiated cell types.

(R5) A differentiation stimulus combined with a type-specific stimulus, both of sufficiently high values and of sufficiently long (finite) durations, should be sufficient to induce differentiation from the progenitor to the osteogenic or respectively chondrogenic cell type.

(R6) The cell type-specific stimuli alone should be insufficient to induce differentiation.

(R7) The interaction network of the model should be weakly connected in a graph theoretical sense, since it is assumed that the TRs of the cell fate switch are not isolated but interacting.

We will refer to these requirements throughout the development and analysis of the model.

2.2.2   Model development

From the requirement (R1), it is already clear that the model will have a three-dimensional state variable, $x = [x_\mathrm{P}, x_\mathrm{O}, x_\mathrm{C}]^T \in \mathbb{R}^3_+$. As a first step, we will choose a plausible network structure.

Network structure   To determine a suitable network structure is a challenge on its own. If measurements of gene expressions are available, network inference methods may be applied (see for example Bansal et al. (2007)). However, if no measurements are available, the model specifications regarding steady states can be consulted to select network structures.

Distinguishing only between activating, inhibiting, or no interaction for each pair of TRs, there are already $3^{n^2} = 19683$ possible network structures. All these three-node gene regulatory network structures were inspected with respect to their ability to yield the three stable steady states, as required in (R2), in Breindl et al. (2011) by applying a quite general approach (Breindl et al., 2010; Chaves et al., 2010). This yielded 206 candidate networks, which were all assembled from building blocks as shown in Figure 2.2 (adopted from Breindl et al. (2011)). From this selection of networks, we will now proceed narrowing down our network choice by excluding building blocks that either vulnerate one of the model requirements (R1)–(R7), or that are less likely than others due to certain arguments.

The first building block in Figure 2.2 represents an isolated TR, which vulnerates the requirement (R7). The sixth building block shows a TR without auto-feedback, which would

Figure 2.2: (adopted from Breindl et al. (2011)) Building blocks from which all three-node networks can be constructed that meet the requirement (R2), to have three stable steady states representing the three cell types, as selected by Breindl et al. (2011). Interaction links that enter x_i from bottom left or bottom right represent interaction links from another x_k, $k \neq i$ and x_l, $l \neq i \wedge l \neq k$. Positive interactions (activation) are denoted by pointed arrows ($\rightarrow$), negative interactions (inhibition) are denoted by blunt arrows ($\dashv$).

contradict the findings that established key TRs of the progenitor, osteogenic, and chondrogenic cell activate their own expression (Drissi et al., 2000; Kumar and Lassar, 2009; MacArthur et al., 2008; Phimphilai et al., 2006). The fifth building block shows a TR that has a negative feedback on its own expression (in addition to the degradation), which may produce oscillating dynamics (Thomas, 1981). This on the one hand does not match with the notion of cell types as stable steady states (R2), and on the other hand again contradicts the findings of positive auto-feedback for these TRs. With this, we are left with a reduced number of candiate networks that can be constructed from the intermediate three building blocks in Figure 2.2 (second, third, fourth building block). These candidate networks are all (except for rotation symmetries) shown in Figure 2.3, in total providing 3+3+1 = 7 different network structures. The thereby obtained networks were found to meet the steady state requirements irrespective of whether the interactions were realized via AND or OR logics[3], a robustness property which further supports the selection of these networks.

From these remaining network structures (Figure 2.3), we choose the first one (without rotation; see Figure 2.4) due to the following reasons: It contains interactions from the progenitor TR, x_P, to both the osteogenic and chondrogenic TR, x_O and x_C, which realizes the assumed influence of the progenitor TR on the cell type-specific TRs and the differentiation stimulus, (R4). It does however not contain any interations from the osteogenic nor chondrogenic TR, x_O and x_C, to the progenitor TR, x_P – such an interaction (from x_O or x_C to x_P) of notable strength may result in a suppression of x_P by solely applying a type-specific input, u_O or u_C, without the need for a differentiation stimulus, u_D, which would contradict (R6).

The chosen model structure consists of two switch parts, each reflecting a different mechanism in the cell differentiation system: A pre-switch determines whether the cell is able to differentiate at all, or remains in the stable progenitor state. Then, the configuration of a fate switch in turn determines into which cell type the cell will differentiate.

Inputs According to (R3), and incorporating (R4), (R5), stimuli enter the system via the following three input components: The unspecific differentiation stimulus u_D suppresses the progenitor TR, whereas the two stimuli acting in the pro-osteogenic or pro-chondrogenic direction, u_O and u_C, enhance the activity level of the cell type-specific TR, respectively. The inputs are depicted within the model structure in Figure 2.4.

---

[3]joint work with C. Breindl, results not shown

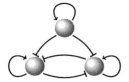 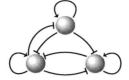 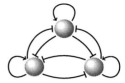

Figure 2.3: Network structures remaining from all the networks constructed from the building blocks shown in Figure 2.2, after imposing the additional selection criterion (R7) and excluding negative auto-feedbacks. Only one representative network is shown for each group of rotation-symmetric networks.

At this point let us note that the components of the input $u = [u_\mathrm{D}, u_\mathrm{O}, u_\mathrm{C}]$ are to be seen as "effective" inputs, rather than the amount of the applied biochemical substance itself. For example, if the cell is stimulated by a substance L which is a ligand, there will be further signal processing through pathways until it finally reaches the nucleus where it affects the TRs (gene expression). Thus, the input signal can be seen as a function $u = u([\mathrm{L}])$ where [L] is the concentration of the substance L. In many cases, the relationship between the administered substance concentration [L] and the effective input u may be plausibly described by a saturating function such as Michaelis-Menten kinetics (Klipp et al., 2000), even without knowing the exact function $u([\mathrm{L}])$.

Kinetic functions After having specified the functional relationships between the model's state variables and inputs, as represented in Figure 2.4, we need to specify kinetic functions in order to derive a system of ODEs for the desired model. Let us rewrite the dynamics $f_i(x, u)$ of each TR from (2.1) as

$$f_i(x, u) = -k_i x_i + \mathrm{act}_i(x). \tag{2.10}$$

The first term on the right hand side represents a linear degradation with degradation rate $k_i \in \mathbb{R}_{++}$, that is a first-order decay, whereas the second term denotes an activation term to be specified. Since the considered gene products are not known to be involved in further higher order degradation processes, this is a suitable choice of kinetic functions, which is commonly used in gene switch models (some examples are Cinquin and Demongeot (2002); Gardner et al. (2000); Glauche et al. (2010); Huang et al. (2007); Laslo et al. (2006); Zhou et al. (2011)).

The activation of each TR's expression arises from the interactions outlined above and depicted in the network structure in Figure 2.4. For each TR $x_i, i \in \{\mathrm{P}, \mathrm{O}, \mathrm{C}\}$, we choose an overall activation term of the Hill-type form

$$\mathrm{act}_i(x) := \frac{b_i + \displaystyle\sum_{\{j \,|\, x_j \to x_i\}} a_{ij} x_j^{n_{ij}} + \displaystyle\sum_{\{j \,|\, u_j \to x_i\}} u_j}{m_i + \displaystyle\sum_{\{j \,|\, x_j \to x_i\}} c_{ij} x_j^{n_{ij}} + \displaystyle\sum_{\{j \,|\, x_j \dashv x_i\}} c_{ij} x_j^{n_{ij}} + \displaystyle\sum_{\{j \,|\, u_j \dashv x_i\}} u_j}. \tag{2.11}$$

In contrast to multiplicative or additive activation terms (Glauche et al., 2010; Huang et al., 2007; Laslo et al., 2006), this formulation imposes no particular assumptions on dimerization

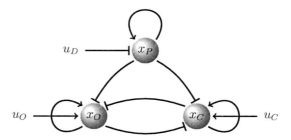

Figure 2.4: The network structure of the model for the gene switch, showing interactions that will be implemented in the model equations. Degradations of each TR are implicitly contained. Positive interactions (activation) are denoted by pointed arrows ($\rightarrow$), negative interactions (inhibition) are denoted by blunt arrows ($\dashv$). Abbreviations: x_P progenitor maintenance TR, x_O osteogenic TR, x_C chondrogenic TR, u_D differentiation stimulus, u_O osteogenic stimulus, u_C chondrogenic stimulus.

or higher order interactions between the TRs and is also used in other publications (Cinquin and Demongeot, 2002; Zhou et al., 2011).

The parameter $b_i \in \mathbb{R}_+$ stands for the basal expression activity of the TR x_i, whereas $m_i \in \mathbb{R}_{++}$ gives the inflection point of the Hill curve. The remaining terms in the nominator arise from all TRs $x_j \rightarrow x_i$ that have an activating interaction on x_i, with activation parameters $a_{ij} \in \mathbb{R}_+$. Since we only have auto-activating interactions in our model structure, that is for all $i: a_{ij} = a_{ii}$, we will write $a_i := a_{ii}$ in the remainder for short. All interactions between TRs have a Hill coefficient $n_{ij} \in \mathbb{N}$. For simplicity, we will assume the same Hill coefficient for all interactions, $\forall i,j: n_{ij} = n$. The last term of the nominator represents inputs $u_j \rightarrow x_i$ that have an activating influence on x_i, with input values $u_j \in \mathbb{R}_+$.

In the denominator, the first summand arises again from all TRs $x_j \rightarrow x_i$ that have activating interactions, in order to impose a saturation of the activations, with inhibition parameters $c_{ij} \in \mathbb{R}_+$. The assumption that any activation between TRs is limited to a certain saturation level is reasonable, since besides the activating TR further resources such as enzymes are usually required (Klipp et al., 2000). The second summand then captures all inhibiting interactions from $x_j \dashv x_i$, with inhibition parameters $c_{ij} \in \mathbb{R}_+$. The last term of the denominator represents inputs $u_j \dashv x_i$ that have an inhibiting influence on x_i, with input values $u_j \in \mathbb{R}_+$.

With this, the functional relationships can be summarized in a set of ODEs,

$$\dot{x}_\mathrm{P} = -k_\mathrm{P}x_\mathrm{P} + \frac{a_\mathrm{P}x_\mathrm{P}^n + b_\mathrm{P}}{m_\mathrm{P} + u_\mathrm{D} + c_\mathrm{PP}x_\mathrm{P}^n}, \tag{2.12}$$

$$\dot{x}_\mathrm{O} = -k_\mathrm{O}x_\mathrm{O} + \frac{a_\mathrm{O}x_\mathrm{O}^n + b_\mathrm{O} + u_\mathrm{O}}{m_\mathrm{O} + c_\mathrm{OO}x_\mathrm{O}^n + c_\mathrm{OC}x_\mathrm{C}^n + c_\mathrm{OP}x_\mathrm{P}^n}, \tag{2.13}$$

$$\dot{x}_\mathrm{C} = -k_\mathrm{C}x_\mathrm{C} + \frac{a_\mathrm{C}x_\mathrm{C}^n + b_\mathrm{C} + u_\mathrm{C}}{m_\mathrm{C} + c_\mathrm{CC}x_\mathrm{C}^n + c_\mathrm{CO}x_\mathrm{O}^n + c_\mathrm{CP}x_\mathrm{P}^n}, \tag{2.14}$$

with initial conditions $x(0) = x_\mathrm{ini}$, and for $i,j \in \{\mathrm{P,O,C}\}$ state variables $x_i \in \mathbb{R}_+$, and parameters $k_i, a_i, b_i, c_{ij} \in \mathbb{R}_+, m_i \in \mathbb{R}_{++}, n \geq 2$. The meaning of the parameters is provided

Parameter	Meaning	Value
n	Hill coefficient	2
a_P	auto-activation of x_P	0.2
b_P	basal activity of x_P	0.5
m_P	inflection point for x_P	10
c_{PP}	self-inhibition strength of x_P	0.1
k_P	decay rate of x_P	0.1
a_O, a_C	auto-activation of x_O, x_C	0.1
b_O, b_C	basal activity of x_O, x_C	1
m_O, m_C	inflection point for x_O, x_C	1
c_{OO}, c_{CC}	self-inhibition strength of x_O, x_C	0.1
c_{OC}, c_{CO}	mutual inhibition strength of x_O, x_C	0.1
c_{OP}, c_{CP}	inhibition strength of x_P on x_O, x_C	0.5
k_O, k_C	decay rates of x_O, x_C	0.1

Table 2.1: Parameter set used for bifurcation analyses and simulations. The parameters are without units.

in Table 2.1, along with the choice of parameter values to be specified next. The inputs u_i for $i \in \{D, O, C\}$ are each $u_i \in [0, u_i^{max}]$. The input u_D corresponds to a (type-unspecific) pro-differentiation stimulus, whereas u_O and u_C to pro-osteogenic and pro-chondrogenic stimuli, respectively. The model could be simplified by reducing the number of parameters (as in Appendix A), but here we use the unreduced equations as the most general representation. Now, the parameter values will be specified.

Parameters Since the model will be generic in the sense that it rather gives a concept of functional relationships, we introduce several simplifying assumptions. First, the parameters in the fate switch (2.13), (2.14) are assigned symmetric values, unless stated otherwise. That means that there is no inherent bias of the cell towards one or the other lineage, as long as no cell type-specific stimulus is applied. Second, the Hill coefficient is chosen $n = 2$ which is the lowest value producing a sigmoidal shape of the activation term. Qualitative results therefore apply equivalently for higher Hill coefficients $n > 2$, arising from more complex reactions.

Further parameter values are now chosen such that the specified model (2.12)–(2.14) fulfills the requirement of three stable steady states (R2). These nominal values of the parameters along with their meaning are summarized in Table 2.1. The model requirements (R2), (R5), (R6) are indeed not only fulfilled for this exemplary choice, but for a certain range of parameter values. We prove this for (2.12) in Appendix B, whereas for (2.13), (2.14) this was ensured via extensive bifurcation analysis (detailed results not shown here).

2.2.3 Model analysis

We analyze the developed model with the aim to verify its ability to reproduce the required properties (R1)–(R7), and to illustrate the biological system in terms of the stability and bifurcation properties of the model. Therefore, we study the stable steady states via multistability analysis, as well as the transitions between stable steady states via bifurcation analysis.

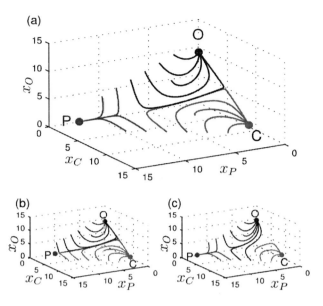

Figure 2.5: Steady states and exemplary trajectories for (a) $u_D = u_O = u_C = 0$, (b) $u_D = 0.5$, (c) $u_O = 0.5$ (other inputs $u_i = C$ unless denoted differently). The three stable steady states are marked with circles and the corresponding cell type P, O, C. The attractor basins Ω_j are visualized by trajectories $x(t, x_0)$ for exemplary starting points x_0. Inputs affect the attractor basins as follows: (b) u_D decreases Ω_P, and increases Ω_O and Ω_C equally. (c) u_O increases Ω_O and decreases Ω_C. Parameters are as in Table 2.1.

Multistability analysis The stability properties are analyzed with respect to the capability of the model to reflect the cell types of the system: the progenitor P, the osteogenic cell type O, and the chondrogenic cell type C, as characterized by (2.6).

The model indeed is able to generate exactly three stable steady states. Figure 2.5a illustrates the three-dimensional state space for an exemplary parameter set as in Table 2.1. Comparing Figure 2.5a against Figure 2.1, one perceives three stable steady states with correct locations in the state space. The existence of further stable steady states was ruled out by exhaustive simulation studies. As required, the three stable steady states represent the states P, O, C and each satisfies (2.6).

Furthermore, each steady state $x^{*,j}, j \in \{P, O, C\}$, is enclosed by an attractor basin (2.4). The extent of an attractor basin, bordered by separatrices, thereby has an actual meaning, in terms of qualitative system properties as well as in the biological context: How close the separatrix is to a stable steady state is one determinant of how robust the system is to fluctuations, for example in mRNA production of the respective TR, and to perturbations, for example by stimuli. A stimulus input can cause a stable steady state to vanish and thus induce a transition to another stable steady state. This will be the focus of the next paragraph.

Bifurcation analysis   Besides the existence of stable steady states, of same interest and importance are the transitions between them as they were outlined in Figure 2.1 and (R5). The differentiation transitions Δ^{diff} as in (2.7) should be achievable by transient system inputs, which correspond to stimuli in the biological language, while dedifferentiation (2.8) and transdifferentiation (2.9) should not be possible. To evaluate whether the model captures the required transitions, we investigate the effect of inputs on the steady states via bifurcation analysis (conducted via the software package CL_MatCont, Dhooge et al. (2003)). Due to the symmetries between (2.13) and (2.14) it is sufficient to investigate only u_O. Equivalent conclusions will then hold for u_C.

The bifurcation analysis reveals that for the parameter set given in Table 2.1, the model allows for exactly the required transitions Δ^{diff}, while excluding Δ^{tra} and Δ^{de} as demanded: The inputs u_D and u_O affect the attractor basins and can shift the system from a tristable to a bistable regime, as illustrated in Figure 2.6. In this case, coming from the state that is lost upon the input signal, the system will converge to one of the two remaining stable steady states. The input u_D decreases Ω_P (Figure 2.5b) and can switch off the pre-switch mechanism for $u_D > u_D^{\text{crit}}$, forcing the system to leave the P state and triggering a differentiation transition Δ^{diff}.

In contrast, the input u_O can not cause the P state to vanish and thus can not induce differentiation (Δ^{diff}) by itself. Instead, u_O enlarges Ω_O (Figure 2.5c). As can be seen in Figure 2.6b, for a sufficiently high value of the osteogenic stimulus $u_O > u_O^{\text{crit}}$ the system could switch from state C to state O, which would mean a trans-differentiation Δ^{tra}. Since this behavior is not observed in the considered cell system, the model might give the hint that the effective input u_O, as applied in the experiment, is limited to a $u_O^{\text{max}} < u_O^{\text{crit}}$. Equivalent results hold for u_C.

To sum up our results from stability and bifurcation analysis, we have shown that the model we propose is able to capture the fundamental properties of the cell differentiation process. This means that it can reproduce the observed number of stable steady states, as well as the transitions between them. As emphasized before, these model properties not only hold for the exemplary parameter set in Table 2.1, but for a certain range of parameters.

2.2.4   Simulations of single cell dynamics

After a general analysis from a theoretic viewpoint in the previous section, we now demonstrate how our model can also serve for more descriptive applications to the biological problem. Based on our model we present two kinds of simulations: In this subsection, the modeling of single cell dynamics serves to investigate the effects of deterministic or extrinsic signals, usually applied on purpose in the form of biochemical stimuli. Later in Section 2.3, simulations on a cell population level will enable also a better understanding of intrinsic signals, given by heterogeneous or stochastic intracellular properties.

We now demonstrate that our model can reproduce the differentiation process upon extrinsic signals on a single cell level. In particular, we study stimuli $u_i, i \in \{D, O, C\}$, or combinations thereof, potent to induce an osteogenic differentiation transition $(P \rightarrow O) \in \Delta^{\text{diff}}$. For simplicity and since the exact functions $u_i([L])$ are unknown, we consider only step function inputs

$$u_i(t) = \begin{cases} \bar{u}_i & \text{for } t_i^{(1)} \leq t \leq t_i^{(2)} \\ 0 & \text{otherwise.} \end{cases} \tag{2.15}$$

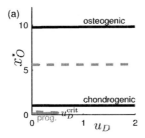 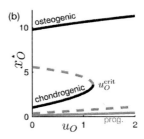

Figure 2.6: Bifurcation diagrams for x_O: (a) vs. the input u_D, and (b) vs. the input u_O. Stable steady state manifolds are given as solid lines, unstable steady state manifolds as dashed lines. The grey solid line corresponds to the P state where x_P is high and thus suppresses x_O generally. The black solid lines correspond to the differentiated states O, C. Parameters are as in Table 2.1. (a) For a sufficiently high input $u_D > u_D^{crit}$, the system is shifted from a tristable to a bistable regime (Ω_P vanishes), and will either converge to the O or the C stable steady state. After removing the input u_D this differentiated state (osteogenic or chondrogenic) is maintained irreversibly. (b) The input u_O can be viewed as acting in a dimension orthogonal to u_D (cf. (a)). Applying u_O will increase Ω_O, until for $u_O > u_O^{crit}$ loosing the C state (Ω_C vanishes), which would correspond to trans-differentiation. After removing the input u_O the state O is maintained irreversibly.

In order to achieve osteogenic differentiation (P $\rightarrow$ O), we know from the previous Subsection 2.2.3 that two goals have to be pursued in order to achieve the desired differentiation to the osteogenic cell type:

1. The system has to escape the attractor basin of P.

2. The system has to enter the attractor basin of O.

Since u_C does not assist in achieving any of these two goals, we consider combinations of u_D and u_O for the following stimulus scenarios:

(a) u_D, u_O are applied concomitantly: $t_D^{(1)} = t_O^{(1)} = 100$.

(b) u_D is applied first, then u_O is applied in addition: $t_D^{(1)} = 100, t_O^{(1)} = 500$.

(c) u_O is applied first, then u_D is applied in addition: $t_D^{(1)} = 500, t_O^{(1)} = 100$.

In order to investigate the effect of asymmetries in parameters (meaning an intrinsic bias towards one of the two lineages), we furthermore consider the following scenario:

(d) u_D is applied first, then u_O is applied in addition: $t_D^{(1)} = 100, t_O^{(1)} = 500$; but (in contrast to scenario (b)) now with parameters $b_O = 0.9 < b_C = 1$, which means a slight intrinsic bias towards chondrogenic lineage; whereas the remaining parameters are as in Table 2.1.

For all scenarios, we set always $\bar{u}_D = 0.8 > u_D^{crit}, \bar{u}_O = 1 < u_O^{crit}, t_D^{(2)} = t_O^{(2)} = 800$, and the remaining parameters as in Table 2.1.

The outcomes of the scenarios (a)–(d) are depicted in Figure 2.7. We will discuss the observations: In scenario (a), the concomitant application of u_D and u_O induces escape from the progenitor state P and attraction to the osteogenic state O (Figure 2.7a). This is also what one expects from the model analysis (Subsection 2.2.3).

In scenario (b), first u_D causes x_P to decrease, thus both x_O and x_C increase to an equal value (Figure 2.7b); but only after u_O is added, x_O fully increases and x_C again decreases to a low value, corresponding to state O. This can be understood by recalling the results from the model analysis: When first only u_D is applied, it temporarily shifts the system from a tristable to a bistable regime (recall Figure 2.6a), causing the system to escape the P state. Since this scenario assumes symmetry of the parameters in (2.13) and (2.14), and in initial values of x_O and x_C as well, the system state will preliminarily remain at a point on the separatrix between the two differentiated stable states (confer Figure 2.5). Then, adding u_O finally pushes the system into the O attractor basin Ω_O.

It is of course important to note that any asymmetries in parameters, initial values, or stochastic fluctuations could cause a deviation from the unstable state and thus attraction to the O or C state, depending on the particular deviation. As a consequence, the eventual cell fate completely eludes controlled guidance. In order to avoid this case, a reasonable guideline would be to never apply a stimulus u_D alone before the additional application of the cell type-specific stimulus u_O. Indeed, to the best of the author's knowledge, protocols for differentiation experiments stipulate to apply differentiation stimuli (as may be ascorbic acid or dexamethasone) and cell-type specific stimuli (such as β-glycerophosphate for osteogenic differentiation of MSCs) concomitantly.

In scenario (c) the two stimuli are applied in the opposite order, thus avoiding the problem just discussed. As seen in Figure 2.7c, first applying u_O slightly increases x_O, but can not cause the P state to vanish (recall Figure 2.6b). In terms of Figure 2.5, this means that the P state is shifted slightly towards O, but the system is still trapped in the attractor basin Ω_P. Only after u_D is added, the tristable regime is left to bistability, where due to the effect of u_O the system enters Ω_O. In this scenario, the differentiation process is completely under control. (In Subsection 2.3.2, we will see that further limitations on this control might be imposed, by considering stochastic effects.)

The symmetry in the parameters assumed so far represents a special case. For the general asymmetric case, we now investigate scenario (d) which reflects an intrinsic bias towards the chondrogenic lineage. As seen in Figure 2.7d, upon applying u_D and loosing the P state the system falls into Ω_C and thus x_C increases to $x_C > x_O$. The application of u_O can not re-attract the system to Ω_O any more: The system is already trapped too deeply in the attractor basin of the chondrogenic state, Ω_C, and since in the scenario u_O is limited to $u_O^{max} < u_O^{crit}$, there exists a certain time point ($t \leq t_O^{(1)} = 500$ here) from which on the input u_O is not sufficient any more for the system to escape Ω_C. This again confirms that asymmetries in parameters can play a crucial role, especially if an unspecific stimulus u_D is to exhibit its effect before the osteogenic stimulus u_O, as in scenario (b).

To sum up the observed dynamics in the simulation scenarios, we can restrict stimulus combinations for successful osteogenic differentiation as follows:

1. There has to be a sufficiently large $u_D > u_D^{crit}$ (confer Figure 2.6a) in order to leave Ω_P.

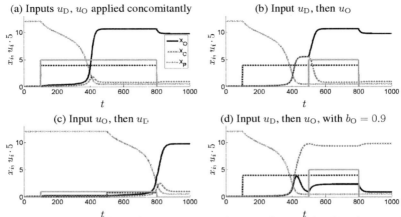

Figure 2.7: Dynamics of stimulus scenarios (a), (b), (c), (d), as outlined in the text. TR dynamics of x_P (black -.), x_O (blue -), x_C (green - -). Stimulus inputs u_D (red thin - -), u_O (cyan thin -) are depicted with factor 5 for clear visualization.

2. As soon as the attractor basin of P is left, and any deviation from the separatrix occurs, for example, due to initial state values asymmetric between x_O and x_C, the system will converge to O or C. There has to be an osteogenic stimulus u_O in order to safely guide the system to Ω_O.

3. Asymmetries in parameters or initial values may impose a bias towards one or the other cell type. In case of a bias towards the undesired state C, the osteogenic stimulus u_O has to be sufficiently large in order to overcome this bias and instead guide the system into Ω_O.

4. The inputs have to be present long enough and at the crucial time points: The differentiation stimulus u_D has to act sufficiently long for x_P to leave Ω_P in order to achieve irreversible escape from the progenitor state. The cell type-specific stimulus u_O has to act sufficiently long to drive the system into the desired Ω_O.

As seen, not only the combination and amount of stimuli is crucial, but also their sequence and times at which they become effective.

None of the two inputs u_D, u_O alone is sufficient to execute a transition guided towards one specific cell type O or C. So in order to steer the system on purpose towards the desired (say, osteogenic O) state, a combination of both inputs is necessary. Biologically, combining the two inputs in any desirable way requires that u_D, u_O could be achieved by biochemically distinct substances and pathways. Indeed, classical osteogenic differentiation medium consists of several biochemical substances (for example, dexamethasone, ascorbic acid, and β-glycerophosphate (Shui et al., 2003)). As long as the details of pathways futher involved in the transmission of these stimuli are not known, it may also be the case that the two effective inputs result from intertwined pathways, in which case a stimulus protocol as suggested by the model may not be possible. Our model can serve to apply hypothetical stimulus combinations and sequences, and compare the resulting outcomes.

2.3 Stochastic dynamics on cell population level

The differentiation scenarios examined so far were all based on a deterministic single cell model, such that the system response to some input signal was precisely as predicted by the bifurcation analysis. But, as already indicated in the previous part 2.2.3, under realistic circumstances the gene expression levels are subject to stochastic fluctuations (Raj and van Oudenaarden, 2008). Such intrinsic signals escape the control via experimental manipulations, and thus can impose limitations on the differentiation success rate. For example, with human bone marrow derived MSCs in osteogenic differentiation experiments typically a success rate of only about $20 - 30\%$ is achieved[4]. It is therefore crucial to identify potential causes and thus improvements.

Since stochastic processes introduce a certain degree of randomness of the trajectory, the resulting dynamics are suitably studied on the cell population level. This section will model the gene switch with additional stochastic fluctuations: First, we specify the stochastic dynamics acting on the gene switch in 2.3.1. Based on this, we extend the ODE model derived beforehand to an SDE model in Subsection 2.3.2.

2.3.1 Specification of stochastic dynamics

The SDE-based Chemical Langevin Equation (2.2) requires the specification of a drift- and a diffusion term. The drift term represents the deterministic dynamics, whereas the diffusion term captures the stochastic dynamics. An ODE model (2.1) can be interpreted as an approximation of a stochastic model if stochastic processes are simply neglected (Higham, 2008). Both have state variables of the same dimension, whereas for the SDE model one needs to rewrite the state variables and inputs into stochastic processes, denoted by X_t and U_t. Thus, for the SDE model the state variables will be $X_t = [X_{P,t}, X_{O,t}, X_{C,t}]^T \in \mathbb{R}^3$ and the input $U_t = [U_{D,t}, U_{O,t}, U_{C,t}]^T \in \mathbb{R}^3$.

Using this relation between ODE and SDE models for our model (2.12)–(2.14), we can directly specify the drift term from (2.10), $f : \mathbb{R}_+^n \times \mathbb{R}_+^m \to \mathbb{R}_+^n$ as

$$f_i(X_t, U_t) = -k_i X_{i,t} + \text{act}(X_t, U_t). \tag{2.16}$$

It remains to specify the diffusion term, which represents the stochastic dynamics. We aim to consider stochastic fluctuations that affect the gene expression dynamics. It is reasonable to assume that mainly the activation terms will be exposed to stochasticity, but not the degradation terms (Pedraza and Paulsson, 2008). With this we get for the diffusion term, $g : \mathbb{R}_+^n \times \mathbb{R}_+^m \to \mathbb{R}_+^n$,

$$g_i(X_t, U_t) = \sigma \sqrt{\text{act}(X_t, U_t)}, \tag{2.17}$$

with the impact of noise given by $\sigma \in \mathbb{R}_+^n$, that is for the here considered model $\sigma = [\sigma_P, \sigma_O, \sigma_C]^T \in \mathbb{R}_+^3$.

---

[4]personal communication with A. Hausser, Institute for Cell Biology and Immunology, University of Stuttgart

Noise is represented by n independent Wiener processes, that is random variables sampled from a standard normal distribution,

$$W_{i,t} \sim \mathcal{N}(w|0,1), \quad i \in \{P, O, C\}. \tag{2.18}$$

This Wiener process determines the term dW_t in the SDE (2.2). For a mathematically founded discussion of the relationship between the Wiener process $W_{i,t}$ and the term $dW_{i,t}$, the reader is referred to Kloeden and Platen (1999).

2.3.2 A stochastic model of the gene switch

We now aim to investigate gene switch dynamics under stochastic fluctuations, and to especially study which additional effects arise that could not be observed from the deterministic model. From the ODE model (2.1) developed for the gene switch, (2.12)–(2.14), we derive an SDE model as given by (2.2).

Derivation of a stochastic differential equation model   With the chosen drift and diffusion term and noise (2.18), one arrives at the SDE model, in the Langevin formalism,

$$dX_P = \left(-k_P X_P + \frac{a_P X_P^n + b_P}{m_P + u_D + c_{PP} X_P^n} \right) dt + \sigma_P \sqrt{\frac{a_P X_P^n + b_P}{m_P + u_D + c_{PP} X_P^n}} dW_{P,t} \tag{2.19}$$

$$dX_O = \left(-k_O X_O + \frac{a_O X_O^n + b_O + u_O}{m_O + c_{OO} X_O^n + c_{OC} X_C^n + c_{OP} X_P^n} \right) dt$$
$$+ \sigma_O \sqrt{\frac{a_O X_O^n + b_O + u_O}{m_O + c_{OO} X_O^n + c_{OC} X_C^n + c_{OP} X_P^n}} dW_{O,t} \tag{2.20}$$

$$dX_C = \left(-k_C X_C dt + \frac{a_C X_C^n + b_C + u_C}{m_C + c_{CC} X_C^n + c_{CO} X_O^n + c_{CP} X_P^n} \right) dt$$
$$+ \sigma_O \sqrt{\frac{a_C X_C^n + b_C + u_C}{m_C + c_{CC} X_C^n + c_{CO} X_O^n + c_{CP} X_P^n}} dW_{C,t} \tag{2.21}$$

with initial conditions $X_{t=0} = X_{\text{ini}}$. This set of stochastic differential equations can be solved numerically via an Euler-Maruyama scheme (Higham, 2001). For the noise parameters let $\sigma_i = 0.4$, such that the state variables X_i experience a coefficient of variation, $\forall i \in \{P, O, C\}$:

$$\frac{\sqrt{\mathbb{E}(X_i^2) - (\mathbb{E}(X_i))^2}}{\mathbb{E}(X_i)} \approx 0.15 \tag{2.22}$$

which is a reasonable magnitude of noise as it is found in comparable systems (Niepel et al., 2009; Pedraza and Paulsson, 2008).

Simulations of stochastic cell population dynamics   We now aim to investigate the dynamics of a population of N cells, each driven by a stochastic gene switch (2.19)–(2.21). This approach of simulating N trajectories of the model equations is also known as ensemble modeling (Stamatakis, 2010), here with stochastic dynamics. As in the deterministic model, stimuli

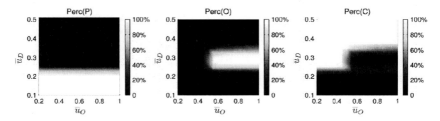

Figure 2.8: The percentages of cells for each cell type, Perc(P), Perc(O), and Perc(C), after stimuli were applied with $u_D(t) = \bar{u}_D$ for $t \in [100, 800]$, $u_O(t) = \bar{u}_O$ for $t \in [400, 800]$ (for details, confer the text). It is seen that the percentage of cells remaining in the progenitor state, Perc(P), is only determined by the differentiation stimulus u_D. The percentage of cells successfully differentiated into the osteogenic cell type, Perc(O), has its maximum for intermediate values of the differentiation stimulus u_D and high values of the osteogenic stimulus u_O. It decreases again for high u_D. The percentage of cells differentiated into the undesired chondrogenic cell type, Perc(C), has its maximum for high u_D, but low u_O. It peaks for intermediate and high values of the differentiation stimulus u_D, regardless of u_O.

are applied as step function inputs (2.15), again $t_D^{(2)} = t_O^{(2)} = 800$. We then investigate the effect of different values of inputs $\bar{u}_D$ and $\bar{u}_O$ on the percentage of cells that end up in state P, O, or C:

$$\text{Perc}(j) := 100\% \cdot \text{Prob}\left(x(t = 1000) \in \Omega_j\right). \tag{2.23}$$

The probability of the system state being within the attractor basin of state i was approximated by using $N = 1000$ simulations of the stochastic model (2.19)–(2.21).

To evaluate how capable a cell type-specific stimulus u_O is to guide the system towards the desired state O, an inherent bias towards the competing state C was created by setting $b_O = 0.5 < b_C = 1$, all other parameters as in Table 2.1. Again, a population of $N = 1000$ cells was simulated for $t \in [0, 1000]$. Stimuli were applied $u_D(t) = \bar{u}_D$ in the time interval $t \in [t_D^{(1)}, t_D^{(2)}] = [100, 800]$ and $u_D(t) = 0$ otherwise, $u_O(t) = \bar{u}_O$ in the time interval $t \in [t_O^{(1)}, t_O^{(2)}] = [400, 800]$ and $u_O(t) = 0$ otherwise.

We investigate the percentages of cells for each cell type in dependence of the values of the applied stimuli, $\bar{u}_D$ and $\bar{u}_O$, as depicted in Figure 2.8: It can be seen that the percentages of cells depend on the amounts of stimuli $\bar{u}_D$ and $\bar{u}_O$ in a highly nonlinear way. The percentage of cells remaining in the progenitor state, Perc(P), is solely determined by $\bar{u}_D$, as expected. It is close to 100% as long as $\bar{u}_D$ is below some threshold, meaning that the stochasticity is not sufficient to induce differentiation on its own. In contrast, it is minimal (namely 0%) above another threshold, although this threshold still fulfills $\bar{u}_D < u_D^{\text{crit}}$. That means that under stochastic dynamics, these values of $\bar{u}_D$ are sufficient to push each cell over the limit point u_D^{crit} at some time instance within $t \in [100, 800]$ during which $\bar{u}_D$ is applied.

The percentage of cells that differentiate successfully into the desired osteogenic cell type, Perc(O), has its maximum for intermediate values of the differentiation stimulus $\bar{u}_D$ and high values of the osteogenic stimulus $\bar{u}_O$. However, Perc(O) decreases again for high values of the differentiation stimulus $\bar{u}_D$, which is an interesting observation: When the unspecific differentiation stimulus $u_D(t)$ is high, many cells leave the progenitor state Ω_P before the cell type-specific stimulus $u_O(t) = u_O$ becomes effective, therefore the cells in many cases follow their inherent bias to the undesired chondrogenic cell type, Ω_C (due to the parameter $b_O = 0.5 < b_C = 1$).

The percentage of cells that differentiate into the undesired chondrogenic cell type, Perc(C), has the maximum value (of 100%) for high values of the differentiation stimulus $\bar{u}_D$, but low osteogenic stimulus $\bar{u}_O$. Moreover, it has two peaks for low and high values $\bar{u}_D$ regardless of $\bar{u}_O$. The first peak for low $\bar{u}_D$ occurs because for a low $u_D(t)$, also $u_O(t)$ can not exhibit its full effect: The differentiation process is then mainly initiated by stochastic fluctuations, rather than the deterministic parameters. The second peak for high $\bar{u}_D$ originates from the same effect responsible for the decrease of Perc(O) for high u_D, namely because many cells leave Ω_P and enter Ω_C before the osteogenic stimulus u_O gets effective.

After comparing the amounts of stimuli, let us also consider the effect of their timing. Therefore, we investigate the delay $(t_O^{(1)} - t_D^{(1)})$ between effective input $\bar{u}_D$ and $\bar{u}_O$, as well as the duration $(t_D^{(2)} - t_D^{(1)})$ of the input $\bar{u}_D$, both with respect to their effect on the change of percentages Perc(i).

Figure 2.9a depicts the effects of the delay $(t_O^{(1)} - t_D^{(1)})$ exemplarily for $u_D = 0.4, u_O = 0.55$ and with the remaining parameter values as before. Clearly, the delay of $\bar{u}_O$ after onset of $\bar{u}_D$ has an effect on the percentage of cells that differentiated into the desired osteogenic cell type, Perc(O), and a reverse effect on the undesired cell type, Perc(C). Thus, it is seen that the timing of the cell type-specific stimulus crucially affects the differentiation success rate. However, the percentage of cells remaining in the progenitor state, Perc(P), is always zero and not affected by the delay of the osteogenic stimulus, as one would expect.

Figure 2.9b depicts the effects of the duration of differentiation stimulus, $(t_D^{(2)} - t_D^{(1)})$, again exemplarily for $u_D = 0.4, u_O = 0.55$ and with the remaining parameter values as before. In this setting, both stimuli are started at the same time, but $t_D^{(2)}$ is varied and with this the duration of the differentiation stimulus $\bar{u}_D$. It is seen that (in contrast to the delay in (a)) the differentiation stimulus duration, $(t_D^{(2)} - t_D^{(1)})$, affects all the percentages of cells remaining in the progenitor state, Perc(P), of cells successfully differentiating into osteogenic cells, Perc(O), and of cells differentiating into the undesired cell types, Perc(C). For a short duration, most cells remain in the progenitor state, thus Perc(P) is highest. With increasing the duration of the differentiation stimulus, less cells remain in the progenitor state, thus Perc(P) decreases to 0, whereas Perc(O) and Perc(C) increases up to a maximum.

Summing up, our stochastic model demonstrates that stochasticity on the one hand can induce differentiation, even if under purely deterministic dynamics differentiation would be impossible. On the other hand, it was seen that stochasticity can also limit the differentiation success rate due to differentiation into the undesired cell type. The studies here indicate that a reduced differentiation success rate may result if

- the differentiation stimulus u_D is too low or also too high (confer Figure 2.8), or/and applied too shortly,

- the cell type-specific stimulus u_O is too low, or/and applied too late after the onset of the differentiation stimulus.

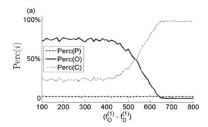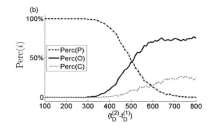

Figure 2.9: The percentages of cells of each cell type, $\mathrm{Perc}(\mathrm{P}), \mathrm{Perc}(\mathrm{O})$ and $\mathrm{Perc}(\mathrm{C})$, versus (a) the delay of the onset of osteogenic stimulus, $(t_\mathrm{O}^{(1)} - t_\mathrm{D}^{(1)})$, (b) the duration of the differentiation stimulus, $(t_\mathrm{D}^{(2)} - t_\mathrm{D}^{(1)})$. (a) The delay between the differentiation stimulus and the osteogenic stimulus, $(t_\mathrm{O}^{(1)} - t_\mathrm{D}^{(1)})$, exhibits a clear influence on the percentages of cells differentiated into the osteogenic cell type, $\mathrm{Perc}(\mathrm{O})$, and into the chondrogenic cell type, $\mathrm{Perc}(\mathrm{C})$. However, the percentage of cells remaining in the progenitor state, $\mathrm{Perc}(\mathrm{P})$, is always zero and not affected by delay of the osteogenic stimulus. (b) The differentiation stimulus duration, $(t_\mathrm{D}^{(2)} - t_\mathrm{D}^{(1)})$, affects the percentages of all three cell types, $\mathrm{Perc}(\mathrm{P})$, $\mathrm{Perc}(\mathrm{O})$, and $\mathrm{Perc}(\mathrm{C})$.

The here presented model of the stochastic gene switch may elucidate causes for observed low percentage of osteogenically differentiated cells, and suggest experimental procedures to increase the differentiation success rate.

2.4 Summary and discussion

In this chapter, we derived and analyzed a deterministic model for a gene switch that drives cell type transitions in stem cell differentiation (Problem 2.1). Furthermore, we derived a stochastic model of the gene switch, appropriate for uncovering dynamics arising on the cell population level (Problem 2.2). Our approach of systematically selecting a network structure and interaction functions of the gene switch, and analyzing the model by systems-theoretic tools, is thereby applicable to similar problems. With this, the work presented in this chapter may serve as a blueprint for the problem of modeling gene switches in cell type transitions.

To approach the problem of model development, we first specified cell types, type-specific gene expression levels, and cell type transitions, as they should be reproduced by the model. In view of these model requirements, the model was then systematically developed: First, we selected network structures that, regardless of the concrete interaction functions, are able to produce the required set of steady states. From these candidate networks, we chose the one with the interaction structure that was rated most plausible based on the biological system properties and the stated requirements. Second, we introduced inputs on the system, representing the impact of biochemical stimuli, also conformly with the model requirements. Third, we built the concrete model dynamics from a general class of kinetic functions, resulting in an unparametrized system of ODEs. Finally, a parametrization of the model was chosen for which it was ensured that the model requirements are fulfilled robustly.

The obtained parametrized deterministic model was then analyzed with respect to its multistability properties and bifurcation behavior. The intention thereby was twofold: The compliance of the model requirements was validated and illustrated. But at least as important, the multistability and bifurcation analysis also yielded a theoretically well-founded understanding of the dynamical properties of cell differentiation. Furthermore, we exploited the model for simulations of distinct scenarios of possible input combinations, and compared the system response. With this we demonstrated how although rather conceptual, a core motif model can already shed light on crucial determinants of cell fate, such as the sequence and timing of stimulus inputs.

To further the understanding of cell fate determination on the gene regulatory level, we advanced the deterministic model to a stochastic model, represented by a system of SDEs. This model was readily obtained from the deterministic dynamics in combination with basic assumptions on the stochastic fluctuations in the dynamics. Since stochastic dynamics can be evaluated suitably on the level of a population of cells, this model was employed for ensemble simulations. These simulations revealed additional limitations for the cell differentiation success rate, such as the magnitude and delay of stimulus inputs.

The chosen approach of ODEs for the deterministic model as well as the choice of SDEs for the stochastic model are thereby of course not the only possible ones. One may well argue that, for example, boolean models might be equally suitable to model single-cell dynamics of gene regulation. They are accompanied by a lower complexity, and are indeed also commonly used to study gene expression or gene switch systems (Kauffman, 1969; Thomas and Kaufman, 2001a,b; Wittmann et al., 2009; Villani et al., 2011). Despite the convenience of a boolean modeling approach, it would also come with several drawbacks for the problems addressed here: Obviously, boolean models do not allow for continuous values of states and inputs, nor of interactions, thereby restricting the system behavior. Similarly, no bifurcation analysis can be conducted for parameters or inputs changing on a continuous scale. Thus, also the effect of changing interaction strengths (parameters) or stimulus amounts (inputs) cannot be studied. Specially suited methods have been dedicated to enable the extension of boolean models to ODE-based models (Krumsiek et al., 2010), as they allow for a higher degree of details.

Contrary to the chosen approach of a core motif gene switch, one may argue that higher-dimensional GRN models would be closer to reality. However, the aim of a core motif model is not to be fit to experimental data and to precisely mimick the quatitative dnamics of complex gene expression networks. Instead, the purpose of such a model is to reproduce conceptual properties, such as steady states, their stability, and switching between them. For this, low-dimensional core motif models have been established as the suitable framework (Chickarmane et al., 2009; Foster et al., 2009; Huang et al., 2007; Roeder and Glauche, 2006), as they are clearly more amenable to systems-theoretic tools like analytical constraints for multistability or bifurcation analysis. For problem settings where a more detailed model is desired, we present a construction method in Chapter 3 to expand a low-dimensional core motif model to a high-dimensional detailed GRN model.

Also for stochastic modeling, further modeling approaches exist such as Markov jump processes (Gillespie, 1977; Higham, 2008), population balance model equations (Ramkrishna, 2000), or the Fokker-Planck equation (Erban et al., 2006; Hasty et al., 1999). It was out of the scope of this chapter to cover the subject of stochastic models and their solution and simulation exhaustively, but refer to, for example, Wilkinson (2006). An alternative approach to model gene switches with stochastic dynamics based on population balance model equations is presented in Chapter 4.

Chapter 3

Construction of multistability-equivalent gene regulatory network models

In this chapter, we present a construction method for multistability-equivalent gene regulatory networks of different dimensionality. Given a low-dimensional dynamical system and the network structure of a more detailed high-dimensional representation of the system, the proposed construction rules yield a high-dimensional dynamical system which exhibits the same multistability properties. In Section 3.1, we first provide some background on models for gene regulatory network dynamics, and then introduce the concept of multistability equivalence in order to formulate the problem addressed in this chapter. In Section 3.2, we present our construction method, and prove that the proposed method indeed yields a multistability-equivalent gene regulatory network model. The construction method is applied in Section 3.3 to investigate a gene regulatory network in the differentiation of mesenchymal stem cells, demonstrating the potential and value of our method. Section 3.4 concludes with a short summary and discussion.

This chapter is based on Schittler et al. (2013b,c, submitted).

3.1 Background and problem formulation

3.1.1 Modeling gene regulatory network dynamics

Gene regulatory networks (GRNs) in biological cells are high-dimensional systems (Bar-Yam et al., 2009; Huang et al., 2005): For example the human genome consists of billions of base pairs, which in turn contain tens of thousands of genes (Alberts et al., 2000). Considering that each gene (or, more precisely, its products) may interact in principal with each other gene, this would yield up to about 10^{70000} interactions, a number beyond any conceivability (Feytmans et al., 2005). Fortunately for the quantitative understanding of gene regulatory networks, by far not all possible interactions are realized, but rather a (relatively small) subset of genes and gene interactions are active. In the human body, for example, about 200 different cell types can be found, and for each one there is a different characteristic set of genes that are actively expressed and interacting (Alberts et al., 2000).

The finding that a cell type's gene regulatory network is made up of a relatively small number of genes is intensified by the identification of a few or even single "master regulator" genes that suffice to characterize the cell fate (Chang et al., 2008; Huang et al., 2007), and of "core motifs" that occur repeatedly in gene regulatory networks and may represent interactions between such master regulators (Alon, 2007a,b; Goldbeter and Koshland, 1984; MacArthur et al., 2009; Tyson et al., 2003; Xiong and Ferrell, 2003). These core motifs are responsible for generating certain properties such as bistability (Gardner et al., 2000; Tyson et al., 2003), switch-like behavior (Cherry and Adler, 2000; Xiong and Ferrell, 2003), oscillations (Tyson et al., 2003), ultrasensitivity (Goldbeter and Koshland, 1984), or adaptivity (Ma et al., 2009; Waldherr et al., 2012), to name some examples reported in the literature. The perception of cell types driven by a relatively small, but nonlinear, dynamical system of interacting genes promoted the study of gene regulatory network dynamics via ordinary differential equation (ODE)-based models that represent abstracted, low-dimensional systems (see, for example, Huang et al. (2007); Li et al. (2006); Narula et al. (2010); Peltier and Schaffer (2010); Roeder and Glauche (2006); Schittler et al. (2010)).

A central property of GRNs is that many of them exhibit multistability, that is the coexistence of several stable states of the system, of which a suitable one can be adopted under the respective biological circumstances (Huang et al., 2007; Narula et al., 2010; Roeder and Glauche, 2006; Schittler et al., 2010; Xiong and Ferrell, 2003). The analysis of multistability properties of such systems is of high interest and has attracted the development of new specially suited methods, for example by Angeli et al. (2004); Breindl and Allgöwer (2009); Breindl et al. (2011); Li et al. (2006); Radde (2010); Thomas (1998); Waldherr and Allgöwer (2009).

It is desired that the analysis of a conceptual, low-dimensional model should allow for conclusions about a higher-dimensional model. On the one hand, low-dimensional systems are amenable to systems theoretic tools such as multistability and bifurcation analysis. On the other hand, a high-dimensional dynamic model might be required to allow for fitting to data from gene expression measurements, and thus for quantitative comparison with and predictions about the real biological system. From this the task arises how a dynamic model for a high-dimensional system can be obtained if only a core motif model of the gene regulatory dynamics, implemented as a low-dimensional ODE system, is available. The high-dimensional ODE system to be obtained should have the same number of steady states, and each steady state should have the same stability properties. This is somewhat complementary to the field of model reduction, where one aims to reduce a high-dimensional model to a low-dimensional one (Antoulas, 2005a,b). Similar questions are also studied for metabolic networks (Conradi et al., 2007), for which the model classes are somewhat different from that for GRNs.

If this question of multistability equivalence could be answered, methods and results from systems analysis of low-dimensional systems and modeling via high-dimensional systems could be combined and thus greatly enhance the understanding of a particular GRN system at hand. The aim of this chapter is to propose construction rules for the expansion of a low-dimensional to a higher-dimensional ODE system, while ensuring multistability equivalence. Thereby, we contribute to the challenge of translating results from core motif models to more realistic, high-dimensional gene regulatory network models.

3.1.2 Problem formulation

In this section we define what we will mean by a multistability-equivalent system, and formulate the problem to be addressed. A low-dimensional system is defined for which the dynamics (and thus also its structure) are given, and a high-dimensional system is defined for which only the structure, but importantly no dynamics, are known.

Let be given an ODE system

$$\Sigma_f : \dot{z} = f(z) = a(z) - d(z), \; z \in \mathbb{R}_+^n, \tag{3.1}$$

with the activation rate $a(z) \in \mathbb{R}_+^n$, and the degradation rate $d(z) = kz$, $k = \text{diag}(k_1, \ldots, k_n)$, with $k_i > 0$.

We assume the system to have R steady states $z^{*(r)}, r = 1 \ldots R$, that is $f(z^{*(r)}) = 0$. The system's Jacobian at a steady state $z^{*(r)}$ is denoted by

$$J_f^{(r)} := \frac{\partial f}{\partial z}(z^{*(r)}). \tag{3.2}$$

The structure of the low-dimensional system will be represented by an interaction sign matrix,

$$S_a := \text{sgn}\left(\frac{\partial a}{\partial z}\right) \in \{-1, 0, +1\}^{n \times n}, \tag{3.3}$$

which is assumed to be constant over $z \in \mathbb{R}_{++}^n$.

The number of unstable modes (positive eigenvalues) for each steady state is given by

$$|\{\lambda_{f,v}^{(r)} | \Re(\lambda_{f,v}^{(r)}) > 0\}|, \tag{3.4}$$

with $\lambda_{f,v}^{(r)}, v = 1, \ldots, n$ the eigenvalues of the Jacobian at the steady state, $J_f^{(r)}$.

Let an interaction sign matrix of a higher-dimensional system be given as

$$S_A \in \{-1, 0, +1\}^{N \times N}, N \geq n. \tag{3.5}$$

The following definition will serve to formulate the concept of a higher-dimensional ODE system

$$\Sigma_F : \dot{x} = F(x) = A(x) - D(x), \; x \in \mathbb{R}_+^N, \tag{3.6}$$

which is desired to have equivalent multistability properties.

Definition 3.1 *A system* (3.6) *is called an N-dimensional multistability-equivalent system to* (3.1) *and consistent with the interaction structure given by* (3.5), *if the following hold.*

(i) *The derivative of the activation function $A(x)$ has signs as given by the interaction sign matrix* (3.5):

$$\text{sgn}\left(\frac{\partial A}{\partial x}\right) = S_A \; \forall x \in \mathbb{R}_{++}^N. \tag{3.7}$$

(ii) *There exists an injective map $h : \mathbb{R}_+^n \to \mathbb{R}_+^N, z^* \mapsto h(z^*)$, with:*

$$f(z^*) = 0 \Leftrightarrow F(h(z^*)) = 0. \tag{3.8}$$

(iii) *The number of unstable modes (positive eigenvalues) in both systems* (3.1) *and* (3.6), *for each pair of steady states $z^{*(r)}$, $x^{*(r)} = h(z^{*(r)})$, is equal:*

$$|\{\lambda_{F,u}^{(r)} | \Re(\lambda_{F,u}^{(r)}) > 0\}| = |\{\lambda_{f,v}^{(r)} | \Re(\lambda_{f,v}^{(r)}) > 0\}|. \tag{3.9}$$

In the remainder, a system that meets the properties of Definition 3.1 will be called *multistability-equivalent* for short.

Let us shortly discuss the motivation behind the chosen definition: The first property (i) states that the constructed system must not have an arbitrary network topology, but that it must fulfill structural properties as far as they can be hypothesized or are even known. In case that the interaction structure is not fully known, several candidate networks may be proposed and the construction will yield accordingly many high-dimensional candidate models. To identify gene regulatory network structures, multiple computational approaches have been developed that use for example gene expression profiles (see for example Bansal et al. (2007) for a review), or that select networks according to multistability properties as provided by Breindl et al. (2011).

By the second property (ii) it is ensured that the constructed high-dimensional system will have a steady state if and only if the low-dimensional system has a steady state. This is important in order to exclude that a construction method for a high-dimensional system may cancel or introduce additional steady states. The last property (iii) requires that not only steady states in both systems should correspond to each other, but also their stability properties. The stability properties are reflected by the real part of the eigenvalues – a nonpositive real part corresponds to a stable mode, whereas a positive real part corresponds to an unstable mode. Therefore, this requirement can be formulated in terms of the number of unstable modes (eigenvalues with positive real part).

Motivated by the need for a rigorous construction method for such a high-dimensional system, as outlined in the previous section, we will now formulate the problem addressed in this chapter.

Problem 3.1 *(Construction of a multistability-equivalent gene regulatory network model) Given a dynamic core motif model (3.1), and the structure of a gene regulatory network via an interaction sign matrix (3.5), construct a dynamic gene regulatory network model (3.6), such that the obtained system (3.6) is multistability-equivalent to the given system (3.1) and consistent with the interaction structure (3.5).*

3.2 Construction method

In order to approach the problem, we first specify in more detail the considered system class for which we aim to find a multistability-equivalent GRN model. Then, we propose a construction method, and prove that this method indeed solves the problem of constructing a multistability-equivalent GRN model as stated beforehand for the considered system class.

3.2.1 Specification of the system class

In this section we specify the system class that will be addressed by the following proposed construction method. The assumptions made on the system class will also be motivated briefly, as they will become even more clear in the detailed proof later.

For the Jacobians $J_f^{(r)}$, $r = 1, \ldots, R$ as in (3.2) it is assumed that they do not have eigenvalues on the imaginary axis. This is important for the structural stability of the system, especially as the argument principle will be exploited later in the proof.

The interaction sign matrix of the low-dimensional system (3.3) is assumed to be constant over $z \in \mathbb{R}^n_{++}$, since otherwise it would depend on the specific values of the state variables z and thus the system structure would be determined not uniquely. The interaction sign matrix of the higher-dimensional system, (3.5) in turn may be available, for example, from qualitative knowledge about gene interactions. These can be represented by an interaction graph, which can be directly translated into a sign matrix (3.5).

The class of considered systems is restricted for technical reasons by two assumptions:

Assumption 3.1 *(Modular structure) Assume that, possibly by reordering the state space variables of system* (3.6)*, the interaction sign matrix* (3.5) *fulfills the following structural property. There exist numbers* $m_i \in \{1, \ldots, n\}$, $i \in \{n+1, \ldots, N\}$, *such that*

- *in rows* $i = n+1, \ldots, N$, *columns* $j = 1, \ldots, n$: $S_A|_{i,m_i} = +1$ *holds for no more than one* m_i.

- *in rows* $i = n+1, \ldots, N$, *columns* $j = n+1, \ldots, N$: $S_A|_{i,j} \in \{0, +1\}$ *holds for all* j *where* $m_i = m_j$, *and* $S_A|_{i,j} = 0$ *otherwise.*

The general structure of such a matrix is given in Fig. 3.1. In the remainder, the first n state variables $\{x_1, \ldots, x_n\}$ will be referred to as "master genes", whereas the remaining $(N-n)$ state variables $\{x_{n+1}, \ldots, x_N\}$ will be referred to as "module genes". In this way, each index $i \in \{n+1, \ldots, N\}$ is uniquely assigned to one interaction module: $i \in \mathcal{M}_{m_i}$, being disjoint subsets of the indices $\{n+1, \ldots, N\} \supseteq \mathcal{M}_j : \bigcap_j \mathcal{M}_j = \emptyset$, and with the assignment such that

- interactions affecting x_i come either from x_{m_i}, or from other x_j belonging to the same interaction module $\mathcal{M}_{m_i}$,

- interactions from x_i go to genes that are master genes or that belong to the same module,

- interactions between genes belonging to the same module are nonnegative.

The classification into master and module genes with an according network structure might be predetermined from biological knowledge. If not, the selection of master genes could be addressed as a separate problem which is beyond the scope of this contribution.

Assumption 3.2 *(Consistency of sign matrices S_a and S_A) For each* $(i,j) \in \{1, \ldots, n\} \times \{1, \ldots, n\}$ *with* $S_a|_{i,j} \neq 0$, *there must exist some simple path from* $j \in \{1, \ldots, n\}$ *to* $i \in \{1, \ldots, n\}$,

$$p_{ij} := ((i, \iota_1), (\iota_1, \iota_2), \ldots, (\iota_{\omega_{ij}-1}, \iota_{\omega_{ij}}), (\iota_{\omega_{ij}}, j))$$
$$\subseteq (\{1, \ldots, N\} \times \{1, \ldots, N\})^{(\omega_{ij}+1)},$$
$$\text{for which} \prod_{(\iota', \iota'') \in p_{ij}} S_A|_{\iota', \iota''} = S_a|_{i,j}. \tag{3.10}$$

Moreover, if there exists $\iota_1 \in \{n+1, \ldots, N\}$ *on some path from* j *to* i: $(i, \iota_1) \in p_{ij}$, *with* $S_A|_{i,\iota_1} \neq 0$, *there has to exist an interaction in the low-dimensional system*

$$S_a|_{i,j} \neq 0. \tag{3.11}$$

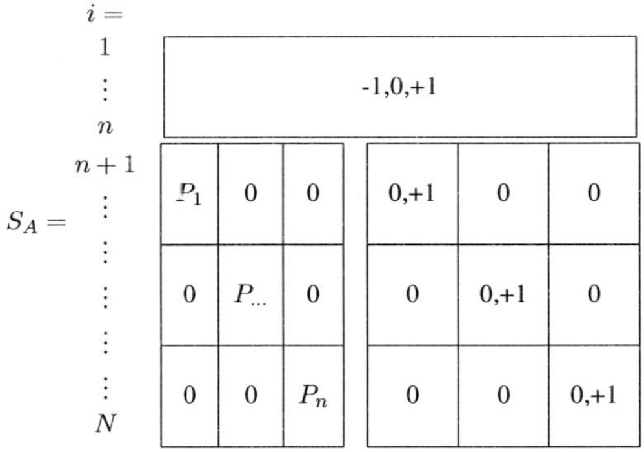

Figure 3.1: General structure of a matrix fulfilling the structural requirements. Each $P_m, m = 1, \ldots, n$ is a column vector of 0 and +1.

If one of these assumptions is not fulfilled, then the proposed construction of an N-dimensional multistability-equivalent system can not be conducted in the way as proposed here, and therefore the existence of an N-dimensional multistability-equivalent system cannot be guaranteed.

3.2.2 Construction rules

In this section we propose a construction procedure for an N-dimensional multistability-equivalent system Σ_F, given the low-dimensional system Σ_f, (3.1), and the high-dimensional interaction sign matrix S_A, (3.5). The idea of the construction is as follows: Additional interactions are introduced via linear activation functions. The remaining interactions are defined in terms of the interactions from the low-dimensional system, with specific mappings between the state spaces of different dimensionality. These interactions are constructed such that the steady state gains of the additional interactions are exactly compensated when the system is at steady state.

The proposed construction procedure is as follows.

Step 1: Construct functions $F_i(x)$ for the module genes indices $i = n+1, \ldots, N$, as follows:

$$F_i(x) = A_i(x) - D_i(x) \quad \text{with}$$
$$A_i(x) = \sum_{j=1,\ldots,N} S_A|_{i,j} x_j, \tag{3.12}$$
$$D_i(x) = K_i x_i, \text{ with } K_i \in \mathbb{R}_{++} \text{ to be chosen.}$$

With this, interaction functions for the $(N - n)$ module genes are determined up to the parameters $K_i, i \in \{n+1, \ldots, N\}$.

Step 2: Next, the influence of master genes on the module genes is captured by defining the following transfer gains. For all (i, k), $k \in \{1, \ldots, n\}$, $i \in \mathcal{M}_k \subseteq \{n + 1, \ldots, N\}$, given a system (3.6), denote the transfer gain with input x_k and output x_i, via the system matrix $(\partial F / \partial x|_{l,m})_{l \in \mathcal{M}_k, m \in \mathcal{M}_k}$, that is, restricted to the state variables in the module $\mathcal{M}_k$:

$$G_{k \to i}(\lambda) = (\mathbf{e}_i^T|_m)_{m \in \mathcal{M}_k} \left(\lambda I_{|\mathcal{M}_k|} - \left(\frac{\partial F}{\partial x}|_{l,m} \right)_{\substack{l \in \mathcal{M}_k \\ m \in \mathcal{M}_k}} \right)^{-1} (S_A|_{l,k})_{l \in \mathcal{M}_k}. \tag{3.13}$$

Then define parameters γ_{ik} which give the corresponding steady state gain for input x_k and output x_i:

$$\gamma_{ik} := G_{k \to i}(0)$$

$$= (\mathbf{e}_i^T|_m)_{m \in \mathcal{M}_k} \left(- \left(\frac{\partial F}{\partial x}|_{l,m} \right)_{\substack{l \in \mathcal{M}_k \\ m \in \mathcal{M}_k}} \right)^{-1} (S_A|_{l,k})_{l \in \mathcal{M}_k}. \tag{3.14}$$

With this step, the signal transmission through the additionally introduced module genes is captured.

Step 3: In this step, functions $F_i(x)$ for the master genes indices $i = 1, \ldots, n$ are constructed.

For the signal transmission between master genes, it is desired to capture which interactions in the high-dimensional system have corresponding interactions in the low-dimensional system with either same or opposite sign. Therefore, for all pairs of genes, (i, ν) for $i, \nu \in \{1, \ldots, n\}$, all interactions in the high-dimensional system from the master gene ν and its module $\mathcal{M}_\nu$ to the master gene i are screened for having the same or opposite sign as the direct interaction from ν to i in the low-dimensional system.

The results are captured in $2n^2$ index vectors, which are each N-dimensional, $J_{(i,\nu)}^=$ and $J_{(i,\nu)}^{\neq}$, that reflect the interactions having the same, respectively the opposite, sign when comparing the high-dimensional to the low-dimensional system's interaction structure. These index vectors $J_{(i,\nu)}^=$ and $J_{(i,\nu)}^{\neq}$, $i, \nu = 1, \ldots, n$, are defined as follows: The j-th entry of $J_{(i,\nu)}^=$ is

$$J_{(i,\nu)}^=|_j := \begin{cases} 0 & , \text{ if } j \notin \mathcal{M}_\nu \cup \{\nu\} \\ |S_a|_{i,\nu}| \cdot \delta(S_A|_{i,j}, S_a|_{i,\nu}) & , \text{ if } j \in \mathcal{M}_\nu \cup \{\nu\} \end{cases}, \tag{3.15}$$

with $\delta(a, b)$ the Kronecker delta of a and b. Similarly, the j-th entry of $J_{(i,\nu)}^{\neq}$ is

$$J_{(i,\nu)}^{\neq}|_j := \begin{cases} 0 & , \text{ if } j \notin \mathcal{M}_\nu \cup \{\nu\} \\ |S_a|_{i,\nu}| \cdot \delta(-S_A|_{i,j}, S_a|_{i,\nu}) & , \text{ if } j \in \mathcal{M}_\nu \cup \{\nu\} \end{cases}, \tag{3.16}$$

These vectors $J_{(i,\nu)}^=$ ($J_{(i,\nu)}^{\neq}$) have an entry 1 in the component j whenever the interaction of z_ν onto z_i in the low-dimensional system has the same (opposite, respectively) sign as the interaction of x_j on x_i in the high-dimensional system, whereas x_j is within the module $\mathcal{M}_\nu$ of x_ν, and zero otherwise.

Furthermore, we define an auxiliary map $\mu_i : \mathbb{R}^N \to \mathbb{R}^n$ to map from the higher-dimensional state space to the lower-dimensional state space. For this purpose, let us define parameters γ_{ik} similarly to (3.14) but for the interactions between master genes, $i, k \in \{1, \ldots, n\}$:

$$\gamma_{ik} := |S_A|_{i,k}|. \tag{3.17}$$

Then, the auxiliary map $\mu_i : \mathbb{R}^N \to \mathbb{R}^n$ is defined as follows:

$$x \mapsto \mu_i(x)$$
$$:= \left(\left(\frac{1 + \epsilon_i \|J_{(i,\nu)}^{\neq}\|}{\|J_{(i,\nu)}^{=}\|} \right) (J_{(i,\nu)}^{=})^T (\gamma_{j\nu}^{-1} x_j)_{j=1,\dots,N} - \frac{\epsilon_i}{\|J_{(i,\nu)}^{=}\|} (J_{(i,\nu)}^{\neq})^T (\gamma_{j\nu}^{-1} x_j)_{j=1,\dots,N} \right)_{\nu=1,\dots,n}$$

(3.18)

with $\gamma_{j\nu} \in \mathbb{R}_{++}$ as determined in (3.14), (3.17). Now, let the functions $F_i(x)$, $i = 1, \dots, n$ be constructed as follows, using the preceding definitions of the index vectors and the auxiliary map:

$$\begin{aligned}
F_i(x) &= A_i(x) - D_i(x) \quad \text{with} \\
A_i(x) &= a_i(\mu_i(x)), \text{ and} \\
D_i(x) &= k_i x_i.
\end{aligned}$$

(3.19)

Step 4: As a last step, the remaining free parameters are chosen. The parameters K_j, for $j \in \{n+1, \dots, N\}$, must be chosen sufficiently large,

$$K_j > K_j^{\min},$$

(3.20)

and the parameters ϵ_i, for $i \in \{1, \dots, n\}$, sufficiently small,

$$0 < \epsilon_i < \epsilon_i^{\max},$$

(3.21)

such that it holds that $\mu_i(x(t)) \in \mathbb{R}_{++}^n$, for all $x \in \mathbb{R}_{++}^N$. It will be proven later in Lemma 3.2 that indeed such $K_j^{\min}$ and $\epsilon_i^{\max}$ exist.

Thereby interaction functions for all $i = 1, \dots, N$ are now determined in terms of function classes. The choice of the free parameters within the constraints, as given in the last step, provides degrees of freedom that can be exploited to, for example, fit the dynamics of a model to data.

3.2.3 Theorem and proof of multistability equivalence

Before we state the theorem that is central for the proposed construction method, we will need the following two lemmata.

The first lemma makes a statement about the system's transfer function for high degradation rates.

Lemma 3.1 *If $K_j = K$ for all $j \in \{n+1, \dots, N\}$, then $\lim_{K \to \infty} \gamma_{ji}^{-1} G_{i \to j}(\lambda) = 1$ for all $i \in \{1, \dots, n\}$, $j \in \mathcal{M}_i$.*

Proof In this case, omitting the subindices and writing $\tilde{F} := \frac{\partial F}{\partial x}$, $\tilde{A} := \frac{\partial A}{\partial x}$, it is

$$\begin{aligned}
\gamma_{ji}^{-1} G_{i \to j}(\lambda) &= \left(e_j^T (-\tilde{A} + KI)^{-1} S_A|_{l,k} \right)^{-1} e_j^T (\lambda I - \tilde{A} + KI)^{-1} S_A|_{l,k} \\
&= \frac{K}{\lambda + K} \left(e_j^T (-\frac{\tilde{A}}{K} + I)^{-1} S_A|_{l,k} \right)^{-1} e_j^T (\frac{-\tilde{A}}{\lambda + K} + I)^{-1} S_A|_{l,k},
\end{aligned}$$

(3.22)

which is equal to 1 in the limit $K \to \infty$. $\qquad\square$

Thereby Lemma 3.1 states that the transfer function value and the steady state gain will cancel each other for high degradation rates, regardless of the value of the transfer function's argument λ. The second lemma makes a statement about the minimum (maximum) values for the parameters K_j (ϵ_i, respectively), as they are demanded in Step 4 in (3.20), (3.21).

Lemma 3.2 *There exist $K_j^{\min} > 0 \, \forall j \in \{n+1, \ldots, N\}$, $\epsilon_i^{\max} > 0 \, \forall i \in \{1, \ldots, n\}$, such that for all $K_j > K_j^{\min}$ and $0 < \epsilon_i < \epsilon_i^{\max}$ it is $\mu_i(x(t)) \in \mathbb{R}_{++}^n \, \forall x \in \mathbb{R}_{++}^N$.*

Proof For appropriate choices of $K_j^{\min}$ with $K_j > K_j^{\min}$, if x_{k_j} is bounded, then also every x_j, $j \in \mathcal{M}_{k_j}$, is bounded. Thus, for every $t > 0$ there exists some $M_{j,k_j}(t) > 0$ for which: $M_{j,k_j}(t)x_{k_j}(t) \geq x_j(t)$. With finite $M \geq \max_{t>0, j \in \mathcal{M}_{k_j}}\{M_{j,k_j}(t)\}$ such that for all $t > 0$, j_1 it is $J_{i,k}^=|_{j_1} = 1$ which lie on a path p_{ik} as in Assumption 3.2, and $j_2 : J_{(i,k)}^{\neq}|_{j_2} = 1$, it holds $Mx_{j_1}(t) \geq x_{j_2}(t)$. Choosing $\epsilon_i \leq \frac{1}{M} \frac{\|J_{(i,k)}^=\|}{\|J_{(i,k)}^{\neq}\|} \frac{\gamma_{j_1 k}}{\gamma_{j_2 k}}$, it is:

$$
\begin{aligned}
\mu_i(x)|_k &\geq \frac{\|J_{(i\,k)}^=\|}{\|J_{(i\,k)}^=\|} \gamma_{j_1 k} x_{j_1} + \frac{1}{M} \frac{\|J_{(i,k)}^=\|}{\|J_{(i,k)}^{\neq}\|} \frac{\|J_{(i,k)}^{\neq}\|}{\|J_{(i,k)}^=\|} \|J_{(i,k)}^=\| \gamma_{j_1 k} x_{j_1} \\
&\quad - \frac{1}{M} \frac{\|J_{(i,k)}^=\|}{\|J_{(i,k)}^{\neq}\|} \frac{\|J_{(i,k)}^{\neq}\|}{\|J_{(i,k)}^=\|} \frac{\gamma_{j_1 k}}{\gamma_{j_2 k}} \gamma_{j_2 k} x_{j_2} \\
&\geq \gamma_{j_1 k} x_{j_1} + \frac{1}{M} \|J_{(i,k)}^=\| \gamma_{j_1 k} x_{j_1} - \frac{1}{M} \gamma_{j_1 k} M x_{j_1} \\
&= \gamma_{j_1 k} x_{j_1} + \frac{\|J_{(i,k)}^=\|}{M} \gamma_{j_1 k} x_{j_1} - \gamma_{j_1 k} x_{j_1} \\
&= \frac{\|J_{(i\,k)}^=\|}{M} \gamma_{j_1 k} x_{j_1} > 0 \quad , \forall x_{j_1} = x_{j_1}(t), \, t > 0. \qquad \square
\end{aligned}
\tag{3.23}
$$

With this, we have ensured the existence of $K_j^{\min}$ ($\epsilon_i^{\max}$, respectively), and thus that appropriate parameters K_j and ϵ_i in Step 4, (3.20), (3.21), can be found.

We now propose and prove that this construction indeed yields multistability-equivalent systems, by showing that a thereby constructed system has the properties (i)-(iii) in Definition 3.1. This is captured by the following

Theorem 3.1 *If (3.5) fulfills Assumptions 3.1 and 3.2, then for any system (3.6) as defined in Section 3.2.1, there exist $K_j^{\min}$, $j = n+1, \ldots, N$ such that every system constructed by the procedure steps 1-4 with $K_j > K_j^{\min}$, $j = n+1, \ldots, N$ is multistability-equivalent to system (3.1).*

Proof We subsequently prove the properties (i)-(iii) from Definition 3.1.

(i) Firstly, we show that the derivative of the constructed activation function $A(x)$ has signs as given by the interaction sign matrix, (3.5).

For indices of the additionally introduced state variables (module genes), $i \in \{n+1, \ldots, N\}$, this property can be seen directly from the construction in (3.12), thus $\forall i \in \{n+1, \ldots, N\}$

$$
\frac{\partial A_i}{\partial x_j} = S_A|_{i,j}.
\tag{3.24}
$$

For the indices of the remaining state variables (master genes), $i \in \{1, \ldots, n\}$, it holds by construction (3.19) that $A_i(x) = (a_i \circ \mu_i)(x)$. From Lemma 3.2, it follows that $\frac{\partial a_i(z)}{\partial z}|_{z=\mu_i(x)} = (S_a)_{i,1,\ldots,n}$. Furthermore, from (3.18) it follows that for $x \in \mathbb{R}_{++}$

$$
\frac{\partial \mu_i(x)}{\partial x_j}\Big|_\nu = \begin{cases} \frac{1+\epsilon_i\|J^{\neq}_{(i,\nu)}\|}{\|J^{=}_{(i,\nu)}\|}(\gamma_{j\nu}^{-1}) > 0 & , \ S_A|_{i,j} = S_a|_{i,\nu} \\ -\frac{\epsilon_i}{\|J^{=}_{(i,\nu)}\|}(\gamma_{j\nu}^{-1}) < 0 & , \ S_A|_{i,j} = -S_a|_{i,\nu} \\ 0 & , \ \text{otherwise} \end{cases} \tag{3.25}
$$

since for all $j \in \mathcal{M}_\nu$ the steady state gain is always nonnegative, $\gamma_{j\nu} \geq 0$. With this, one can write $\forall i \in \{1, \ldots, n\}$, $j \in \{1, \ldots, N\}$:

$$
\begin{aligned}
\mathrm{sgn}\left(\frac{\partial A_i}{\partial x_j}\right) &= \mathrm{sgn}\left(\frac{\partial a_i(\mu_i(x))}{\partial \mu_i(x)}\right) \cdot \mathrm{sgn}\left(\frac{\partial \mu_i(x)}{\partial x_j}\right) \\
&= S_a|_{i,j} \cdot \begin{cases} +1 & , \ S_a|_{i,j} = S_A|_{i,j} \\ -1 & , \ S_a|_{i,j} = -S_A|_{i,j} \\ 0 & , \ \text{otherwise} \end{cases} \\
&= S_A|_{i,j} \quad .
\end{aligned} \tag{3.26}
$$

With the sign equality (3.7) proven for $i \in \{1, \ldots, n\}$ in (3.26), and for $i \in \{n+1, \ldots, N\}$ in (3.24), the property (i) holds.

(ii) Secondly, we prove that there exists an injective map as required in (3.8). Choosing the map $h : \mathbb{R}^n \to \mathbb{R}^N$ as

$$
h : z^* \mapsto h(z^*) := \left((z_i^*)_{i=1,\ldots,n} \ , \ \sum_{\{k : i \in \{\mathcal{M}_k\}\}} (\gamma_{ik}^{-1} z_k^*)_{i=(n+1)\ldots N} \right)^T \tag{3.27}
$$

it is injective, since it is injective in the first n components.

Now, we prove $f(z^*) = 0 \Rightarrow F(h(z^*)) = 0$ separately for $i \in \{1, \ldots, n\}$ and $i \in \{n+1, \ldots, N\}$.

For $i \in \{1, \ldots, n\}$, by construction (3.19) and with (3.18), (3.27) and the structure of the matrix S_A (Fig. 3.1), it is

$$
\begin{aligned}
\mu_i(h(z^*)) &= \left(\left(\frac{1 + \epsilon_i\|J^{\neq}_{(i,\nu)}\|}{\|J^{=}_{(i,\nu)}\|} \right) (J^{=}_{(i,\nu)})^T (\gamma_{j\nu}^{-1} h(z^*)|_j)_{j=1,\ldots,N} \right. \\
&\quad \left. -\frac{\epsilon_i}{\|J^{=}_{(i,\nu)}\|} (J^{\neq}_{(i,\nu)})^T (\gamma_{j\nu}^{-1} h(z^*)|_j)_{j=1,\ldots,N} \right)_{\nu=1,\ldots,n} \\
&= \left(\frac{\|J^{=}_{(i,\nu)}\|}{\|J^{=}_{(i,\nu)}\|} z_\nu^* + \epsilon_i \frac{\|J^{\neq}_{(i,\nu)}\|}{\|J^{=}_{(i,\nu)}\|} z_\nu^* - \epsilon_i \frac{\|J^{\neq}_{(i,\nu)}\|}{\|J^{=}_{(i,\nu)}\|} z_\nu^* \right)_{\nu=1,\ldots,n} \\
&= \left(\frac{\|J^{=}_{(i,\nu)}\|}{\|J^{=}_{(i,\nu)}\|} z_\nu^* \right)_{\nu=1,\ldots,n} = \left(z_\nu^* \right)_{\nu=1,\ldots,n} = z^*.
\end{aligned} \tag{3.28}
$$

With $h(z^*)|_i = z_i^*$, one arrives at

$$
\begin{aligned}
F_i(h(z^*)) &= a_i(\mu_i(h(z^*))) - k_i h(z^*)|_i \\
&= a_i(z^*) - k_i z_i^* = f(z^*) = 0.
\end{aligned} \tag{3.29}
$$

For the remaining indices $i \in \{n+1, \ldots, N\}$, by construction (3.12) it is

$$
\begin{aligned}
F_i(h(z^*)) &= A_i(h(z^*)) - D_i(h(z^*)) \\
&= \sum_{j=1,\ldots,N} S_A|_{i,j} h(z^*)|_j - K_i h(z^*)|_i \\
&= \sum_{j \in \mathcal{M}_{m_i}} S_A|_{i,j} \frac{\gamma_{jm_j}}{\gamma_{im_i}} z_{m_j}^* - K_i \gamma_{im_i} z_{m_i}^*
\end{aligned}
\tag{3.30}
$$

and since it has to hold $z_{m_i}^* = z_{m_j}^* = z_m^*$ for all $(i,j) : S_A|_{i,j} \neq 0$, it is

$$
\begin{aligned}
F_i(h(z^*)) &= \Big(\sum_{j \in \mathcal{M}_m} S_A|_{i,j} \gamma_{jm} - K_i \gamma_{im} \Big) z_m^* \\
&= \Big(\sum_{\substack{j \in \mathcal{M}_m \\ j \neq i}} S_A|_{i,j} \gamma_{jm} + S_A|_{i,i} \gamma_{im} - K_i \gamma_{im} \Big) z_m^*,
\end{aligned}
\tag{3.31}
$$

which, if divided by the sum of steady state gains of all influencing state variables, is

$$
\begin{aligned}
&\Big(\frac{\sum_{\{j:S_A|_{i,j} \neq 0, j \neq i\}} \gamma_{jm}}{\sum_{\{j:S_A|_{i,j} \neq 0, j \neq i\}} \gamma_{jm}} - K_i \frac{\gamma_{im}}{\sum_{\{j:S_A|_{i,j} \neq 0, j \neq i\}} \gamma_{jm}} + \frac{S_A|_{i,i} \gamma_{im}}{\sum_{\{j:S_A|_{i,j} \neq 0, j \neq i\}} \gamma_{jm}} \Big) z_m^* \\
&= \Big(1 + \frac{S_A|_{i,i}}{K_i - S_A|_{i,i}} - K_i \frac{\gamma_{im}}{K_i - S_A|_{i,i}} \Big) z_m^* = \Big(1 + \frac{S_A|_{i,i} - K_i}{K_i - S_A|_{i,i}} \Big) z_m^* = \Big(1 - 1 \Big) z_m^* = 0.
\end{aligned}
\tag{3.32}
$$

With this, it follows that $f(z^*) = 0 \Rightarrow F(h(z^*)) = 0$.

Finally, the reverse direction $f(z^*) = 0 \Leftarrow F(h(z^*)) = 0$ for $i \in \{1, \ldots, n\}$ follows from the previous part: If $F_i(h(z^*)) = 0$, then $0 = a_i(\mu_i(h(z^*))) - k_i h(z^*)|_i = a_i(z^*) - k_i z_i^* = f_i(z_i^*) = 0$. With this, (3.8) holds.

(iii) As the final step, it remains to be shown that the number of unstable modes (eigenvalues with positive real part) of the high-dimensional system (3.6), if constructed by the steps 1-4, is equal to the number of unstable modes (eigenvalues with positive real part) of the low-dimensional system (3.1).

The idea of the proof is the following: At first, for the stability analysis the systems are linearized at the respective steady state. Then, we perform a loopbreaking on both systems that supplies inputs and outputs while ensuring that there are no unstable pole-zero cancellations. From this, we can now write transfer matrices of these systems. Finally, it is shown that the Nyquist curves of these transfer matrices under mild assumptions are that close to each other that both transfer matrices of the loopbroken systems have the same number of zeros in the right half plane. Thus, it can be shown that the transfer matrices of the two closed systems have the same number of poles in the right half plane. With unstable pole-zero cancellations ruled out by the initial loopbreaking, we can conclude that the Jacobians of both systems have the same number of eigenvalues in the right half plane.

The stability of a system at a steady state can be analyzed via the eigenvalues of its Jacobian, that is of the system linearized at this steady state. Therefore, we consider for system (3.1) the system linearized at a particular steady state z^*,

$$
\tilde{\Sigma}_f : \dot{\tilde{z}} = \tilde{f} \tilde{z} = \tilde{a} \tilde{z} - \tilde{d} \tilde{z},
\tag{3.33}
$$

41

with the matrices

$$\tilde{a} := \frac{\partial a(z)}{\partial z}\bigg|_{z=z^*}$$
$$\tilde{d} := \frac{\partial d(z)}{\partial z}\bigg|_{z=z^*},$$

(3.34)

and for system (3.6) the linearized system

$$\tilde{\Sigma}_F : \dot{\tilde{x}} = \tilde{F}\tilde{x} = \tilde{A}\tilde{x} - \tilde{D}\tilde{x},$$

(3.35)

with

$$\tilde{A} := \frac{\partial A(x)}{\partial x}\bigg|_{x=x^*}$$
$$\tilde{D} := \frac{\partial D(x)}{\partial x}\bigg|_{x=x^*}.$$

(3.36)

Let us now introduce the loopbreaking of both systems, somewhat similar to the approach in Waldherr and Allgöwer (2009). (For details the reader is referred to Waldherr and Allgöwer (2009).) The aim of the loopbreaking is to obtain a system with inputs and outputs, and at the same time to rule out any unstable pole-zero cancellations. The former is required to obtain a transfer matrix, while the latter is important to have every unstable eigenvalue supplied by an unstable pole. Therefore, all loops that may possibly produce such cancellations will be broken: The self-loops in the degradation term $\tilde{d}\tilde{z}$ ($\tilde{D}\tilde{x}$, respectively) yield solely negative poles, and thus will not alter the stability. Thus, we only break the loops in $\tilde{a}\tilde{z}$ ($\tilde{A}\tilde{x}$, respectively). The obtained systems will be referred to as loopbroken systems, to emphasize that they are not really open-loop systems since the degradation self-loops are still maintained.

We perform such a loopbreaking in the low-dimensional system, linearized at a particular steady state, $\tilde{\Sigma}_f^{(r)}$. The corresponding loopbroken system reads

$$\dot{z} = -\tilde{d}\tilde{z} + \tilde{a}u$$
$$y = I_n\tilde{z},$$

(3.37)

with the input $u \in \mathbb{R}_+^n$, the output $y \in \mathbb{R}_+^n$, and the linearized dynamics and input matrices as in (3.34). The term $\tilde{a}u$ reflects the activation function of the system (3.33), but the argument $\tilde{z}$ is replaced by u to realize the loopbreaking. The original linearized system (3.33) is obtained by closing the loop via setting the input equal to the output, $u = y$.

We perform a similar loopbreaking in the high-dimensional system Σ_F, linearized at a particular steady state $x^* = h(z^*)$. This results in a loopbroken system

$$\dot{\tilde{x}} = -\tilde{D}\tilde{x} + \tilde{A}U$$
$$Y = I_N\tilde{x},$$

(3.38)

with the input $U \in \mathbb{R}_+^N$, the output $Y \in \mathbb{R}_+^N$, and the linearized dynamics and input matrices as in (3.36). Here, the term $\tilde{A}U$ reflects the linearized activation function of the original system (3.35), but with $\tilde{x}$ replaced by U to realize the loopbreaking. The original linearized system (3.35) is again obtained by closing the loop via setting the input equal to the output, $U = Y$.

We can now write the transfer matrices for the obtained systems. The transfer matrix of the loopbroken system (3.37) is

$$G_f(\lambda) = I_n(\lambda I_n + \tilde{d})^{-1}\tilde{a} \qquad (3.39)$$

and for eigenvalues λ_f of the corresponding closed-loop system it holds that

$$\det\left(I_n - G_f(\lambda_f)\right) = 0 \qquad (3.40)$$

where we used Waldherr and Allgöwer (2009), Lemma 2.3, and the properties of the transfer matrix of a multi-input multi-output system. Since now we have ensured that all positive eigenvalues (unstable modes) will be detected as positive poles in the transfer matrix of the closed-loop system, that is, as zeros λ_f of $\det(I_n - G_f(\lambda))$, we can exploit $\det(I_n - G_f(\lambda))$ to analyze the unstable modes of the system.

Similarly, the transfer matrix of the second loopbroken system (3.38) is

$$G_F(\lambda) = I_N(\lambda I_N + \tilde{D})^{-1}\tilde{A} \qquad (3.41)$$

and for eigenvalues λ_F of the corresponding closed-loop system it is

$$\det\left(I_N - G_F(\lambda_F)\right) = 0. \qquad (3.42)$$

Let us now put in relationship the winding numbers of the Nyquist curves arising from the two transfer matrices, which allows to deduce the number of unstable poles of the corresponding two closed-loop systems.

We now let $\varepsilon := \min_\omega |\det(I_n - G_f(j\omega))|$ denote the minimum distance of this Nyquist curve from the origin. That is, as long as the Nyquist curve $\det(I_N - G_F(j\omega))$ deviates from the Nyquist curve $\det(I_n - G_f(j\omega))$ less than ε (meaning that it lies within a tube of diameter ε), then both Nyquist curves have the same winding number with respect to the origin. Due to the assumption that the Jacobian of $\tilde{\Sigma}_f$ has no eigenvalues on the imaginary axis, we have $\varepsilon > 0$. Now if for all $j : K_j \to \infty$, then the Nyquist curves approach each other, as will be outlined in the following.

From (3.41), it is seen that the rows $(n+1)\ldots N$ of $G_F(\lambda)$ are

$$\left(G_F(\lambda)|_{i,j}\right)_{\substack{i=(n+1)\ldots N \\ j=1,\ldots,N}} = \mathrm{diag}\left((\lambda + K_i)^{-1}_{i=(n+1)\ldots N}\right) \cdot \left(\tilde{A}|_{i,j}\right)_{\substack{i=(n+1)\ldots N \\ j=1,\ldots,N}}. \qquad (3.43)$$

For letting the degradation rates of the module genes, $\forall i \in \{n+1,\ldots,N\} : K_i = K \to \infty$, each entry of (3.43) becomes

$$G_F(\lambda)|_{i,j} = (\lambda + K_i)^{-1}\tilde{A}|_{i,j} \to 0, \qquad (3.44)$$

by Lemma 3.1. Thus as $K \to \infty$, the expression for the determinant becomes

$$\begin{aligned}
\det(I_N - G_F(j\omega)) &\to \det(I_n - G_f(j\omega)) \cdot \det(I_{N-n}) \\
&= \det(I_n - G_f(j\omega)) \cdot 1.
\end{aligned} \qquad (3.45)$$

With this it remains to investigate the remaining upper left $(n \times n)$ subblock of $G_F(\lambda)$, corresponding to the transmission of signals between master genes. Since I_N and $(\lambda I_N + \tilde{D})^{-1}$ in (3.41) are diagonal matrices, this $(n \times n)$ subblock of $G_F(\lambda)$ only depends on the upper left

$(n \times n)$ subblock of the input matrix $\tilde{A}$. Due to the construction of $A(x)$ in step 3, this upper left $(n \times n)$ subblock of the input matrix $\tilde{A}$ in (3.41) is

$$(\tilde{A})_{\substack{i=1,\dots,n \\ j=1,\dots,n}} = \left(\frac{\partial A_i}{\partial x_j}\Big|_{x^*}\right)_{\substack{i=1,\dots,n \\ j=1,\dots,n}} = \left(\frac{\partial a_i(\mu_i(x))}{\partial \mu_i(x)}\frac{\partial \mu_i(x)}{\partial x_j}\Big|_{x^*}\right)_{\substack{i=1,\dots,n \\ j=1,\dots,n}}. \tag{3.46}$$

Therein, x^* can further be substituted via the map $h : \mathbb{R}^n \to \mathbb{R}^N$, as defined in (3.27) to map the steady states of the low-dimensional system to the steady states of the high-dimensional system. According to its definition (3.27), the map h yields in its first n components, $j \in \{1,\dots,n\}$: $x_j^* = (h(z^*))_j = z_j^*$. Using this, we can further rewrite (3.46) to

$$\begin{aligned} \left(\frac{\partial a_i(\mu_i(x))}{\partial \mu_i(x)}\frac{\partial \mu_i(x)}{\partial x_j}\Big|_{x^*}\right)_{\substack{i=1,\dots,n \\ j=1,\dots,n}} &= \left(\frac{\partial a_i(z)}{\partial z}\Big|_{z^*} \operatorname{sgn}\left(\left|\frac{\partial A(x)}{\partial x_j}\right|\right)\right)_{\substack{i=1,\dots,n \\ j=1,\dots,n}} \\ &= \left(\frac{\partial a_i(z)}{\partial z_j}\Big|_{z^*}\right)_{\substack{i=1,\dots,n \\ j=1,\dots,n}} = \frac{\partial a(z)}{\partial z}\Big|_{z^*} = \tilde{a}. \end{aligned} \tag{3.47}$$

Let us now take this result and the fact that the degradation rates for the master genes, $\forall i \in \{1,\dots,n\} : K_i = k_i$, into the expression for the entries in the transfer matrix, as appearing in (3.45). Then, we see that these entries are

$$G_F(\mathrm{j}\omega)|_{i,j} = (\mathrm{j}\omega + k_i)^{-1}\tilde{a}|_{i,j} = G_f(\mathrm{j}\omega)|_{i,j}. \tag{3.48}$$

Since each entry of the upper $(n \times n)$ subblock of $G_F(\mathrm{j}\omega)|_{i,j}$ equals $G_f(\mathrm{j}\omega)|_{i,j}$, and because of (3.45), also the determinants approach each other for $K \to \infty$:

$$\det(I_N - G_F(\mathrm{j}\omega)) \to \det(I_n - G_f(\mathrm{j}\omega)). \tag{3.49}$$

This in turn implies that there exist $K_j^{\min}$ such that the Nyquist curve of the high-dimensional system, $\det(I_N - G_F(\mathrm{j}\omega))$, lies within a tube of diameter $\varepsilon = \min_\omega |\det(I_n - G_f(\mathrm{j}\omega))|$ around the Nyquist curve of the low-dimensional system for sufficiently large K_j, and then the winding numbers of these curves around the origin are equal. Then, by the argument principle, $\det(I_N - G_F(\mathrm{j}\omega))$ has the same number of zeros in the right half plane as $\det(I_n - G_f(\mathrm{j}\omega))$, and the Jacobians of the two systems $\tilde{\Sigma}_f$ and $\tilde{\Sigma}_F$ have the same number of positive eigenvalues.

With this, it is proven that for sufficiently high K_j the property (iii) holds, which concludes the proof. $\qquad\square$

3.3 Example: Mesenchymal stem cell differentiation

In this section, we employ the proposed method to examine a GRN that determines the differentiation of mesenchymal stem cells into the adipogenic, osteogenic, or chondrogenic cell type. Mesenchymal stem cells are a type of adult stem cells that are characterized by their potential to differentiate into adipocytes (fat cells), osteoblasts (bone cells), and chondrocytes (cartilage cells) (Baksh et al., 2004; Heino and Hentunen, 2008; Nakashima and de Crombrugghe, 2003; Ryoo et al., 2006).

A set of genes and transcription factors have been established as cell type-specific markers for the adipogenic, osteogenic, or chondrogenic lineage, as summarized in Table 3.1. A core

Gene	full or alternative name	cell type-specificity	state variable
CEBPβ	CCAAT-enhancer-binding protein α	adipogenic	z_1, x_1
RUNX2	Runt-related transcription factor 2	osteogenic	z_2, x_2
SOX9	Sry-related HMG box	chondrogenic	z_3, x_3
PPARγ	Peroxisome proliferator-activated receptor γ	adipogenic	x_4
CEBPα	CCAAT-enhancer-binding protein β	adipogenic	x_5
LPL	Lipoprotein lipase	adipogenic	x_6
BGLAP	Osteocalcin	osteogenic	x_7
OSX	Osterix	osteogenic	x_8
SPARC	Osteonectin	osteogenic	x_9

Table 3.1: Genes that are established as adipogenic, osteogenic, or chondrogenic markers respectively, of which the mRNA products have been measured in the considered differentiation experiments of mesenchymal stem cells. The first three genes denote "key regulators" or master genes, whereas the remaining genes are measured as additional cell type-specific markers.

motif model of three key regulator genes that each characterize one of the three considered cell types can be derived on the basis of the dynamical properties of the biological system. The derivation of such a core motif model could be conducted via the model development scheme presented in Chapter 2, which was conducted therein for a similar model. However, such a model neither considers all measured genes, nor can it cope with the higher complexity of the real GRN.

For this purpose, one aims to derive a mathematical model of the detailed GRN which determines mesenchymal stem cell fate. Such a model in turn could allow for predicting gene expression dynamics under certain differentiation stimuli, comparing various differentiation protocols, or classifying the therapeutic potential of mesenchymal stem cells from individual donors. Fortunately, the previously presented construction method offers a solution to derive such a model, based on the developed core motif model.

The aim of the example in this section is the following:

- A low-dimensional GRN model is developed with a parametrization such that it reproduces the observed cell types. This low-dimensional model is used to study generic properties of the cell differentiation process, such as the stability of cell types in dependence of parameters.

- A high-dimensional GRN model will be derived via the construction method proposed in this chapter beforehand, which incorporates a set of genes given by the differentiation experiments.

The obtained high-dimensional model can be fit to the full readout of the experimental data, and can be further exploited to determine for example donor-specific differentiation parameters. Although the availability of the high-dimensional model offers a whole new field of model application, including parameter estimation, donor classification, and experimental design, these topics are out of the scope of this thesis.

Specification of GRNs First, the cell types observed in the MSC system under study are briefly characterized. The cell types under consideration are the following three, and are de-

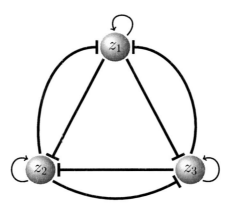

Figure 3.2: Interaction network structure of the low-dimensional GRN, with the state variables representing the master genes as given in Table 3.1. Stump arrows denote negative interactions, sharp arrows denote positive interactions, as given by the interaction sign matrix.

fined according to their expression of type-specific genes as reported in the literature (Baksh et al., 2004; Heino and Hentunen, 2008; Ryoo et al., 2006; Nakashima and de Crombrugghe, 2003; Darlington et al., 1998; Tang et al., 2004; Drissi et al., 2000; Shui et al., 2003; Fu et al., 2007; Zhou et al., 2006; and others):

- Adipogenic cell type: This cell type exhibits a high expression at early stages of CEBPβ, followed by CEBPα, PPARγ, and, at later stages, also LPL.

- Osteogenic cell type: This cell type is characterized by early expression of RUNX2, followed by OSX, BGLAP, and SPARC.

- Chondrogenic cell type: This cell type is characterized by the expression of SOX9.

The cell type-specific genes that have been used in the considered differentiation experiments have been selected as they have been established in the literature, and are also summarized in Table 3.1.

The genes CEBPβ, RUNX2, and SOX9 are established as master regulators in the differentiation of mesenchymal stem cells into adipogenic, osteogenic, and chondrogenic lineage, respectively, and are known to regulate also the expression of the other genes that were selected (confer references above). Based on literature research again, two interaction networks can be drawn: The interaction network of a low-dimensional GRN between the $n = 3$ master genes as depicted in Figure 3.2, and the interaction network of a higher-dimensional GRN between all $N = 9$ genes, as depicted in Figure 3.3. Therein, stump arrows denote negative interactions, sharp arrows denote positive interactions, as defined by entries -1 and $+1$, respectively, in the interaction sign matrix (3.3).

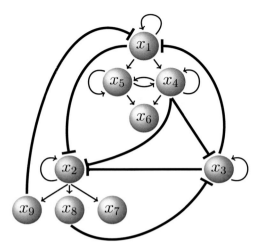

Figure 3.3: Interaction network structure of the high-dimensional GRN, with the state variables representing all genes (master and module genes) as given in Table 3.1. Master genes are denoted in pink, module genes in grey. Stump arrows denote negative interactions, sharp arrows denote positive interactions, as given by the interaction sign matrix.

Let the corresponding dynamical GRN model of the low-dimensional GRN be represented by the following system of ODEs, with the state variables $z \in \mathbb{R}^3$ encoding the expression levels of master genes as summarized in Table 3.1:

$$
\begin{aligned}
\dot{z}_1 &= \frac{0.2z_1^2 + 0.5 + u_A}{10m + 0.1z_1^2 + 0.5z_2^2 + 0.5z_3^2} - 0.1z_1 \\
\dot{z}_2 &= \frac{0.1z_2^2 + 1 + u_O}{1m + 0.1z_2^2 + 0.5z_1^2 + 0.1z_3^2} - 0.1z_2 \\
\dot{z}_3 &= \frac{0.1z_3^2 + 1 + u_C}{1m + 0.1z_3^2 + 0.5z_1^2 + 0.1z_2^2} - 0.1z_3
\end{aligned} \tag{3.50}
$$

From the observations of the biological system, the following system properties are required to be captured by the model:

(R1) There are four free parameters, corresponding to: suppression from stem cell maintenance factors, m, adipogenic stimulus, u_A, osteogenic stimulus, u_O, chondrogenic stimulus, u_C.

(R2) If the system is unstimulated ($u_O = u_A = u_C = 0$) and the stem cell maintenance is low ($m = 1$), the model exhibits three stable steady states, corresponding to the adipogenic, osteogenic, chondrogenic cell type with gene expression levels as described above.

(R3) If an adipogenic (or, osteogenic, chondrogenic) stimulus with sufficiently high value is applied, $u_A > u_A^{crit}$ ($u_O > u_O^{crit}$, $u_C > u_C^{crit}$, respectively), only one stable steady state remains which corresponds to the adipogenic (osteogenic, chondrogenic) cell type.

(R4) If the suppression from the stem cell maintenance factor is kept at a sufficiently high value, $m > m^{crit}$, only one stable steady state remains which corresponds to the mesenchymal stem cell type with low gene expression levels of all type-specific genes.

To derive a core motif model that fulfills this list of requirements, stability and bifurcation analysis can be performed (see the comprehensive model development in Chapter 2). Therefore, we numerically solved the system of equations $\dot{z} = 0$ and conducted bifurcation analysis for the parameters u_A, u_O, u_C, m, respectively. This yields that the requirements (R1)–(R4) are fulfilled for the chosen parameter values as denoted in the following. Exemplarily, bifurcation analysis results are shown for the parameter u_O, the osteogenic stimulus, in Figure 3.4, and will be discussed later.

Construction of high-dimensional GRN   Given the interaction network of the high-dimensional GRN as in Figure 3.3, and the dynamics of the low-dimensional GRN (3.50), the construction procedure presented in Section 3.2 can now be conducted.

The obtained steady state gain parameters are

$$
\gamma_{41} = \frac{K_5}{K_4 K_5 - K_4 - K_5}, \gamma_{51} = \frac{K_4}{K_4 K_5 - K_4 - K_5}, \gamma_{61} = \frac{K_4 + K_5}{K_6(K_4 K_5 - K_4 - K_5)},
$$
$$
\gamma_{72} = \frac{1}{K_7}, \qquad\qquad \gamma_{82} = \frac{1}{K_8}, \qquad\qquad \gamma_{92} = \frac{1}{K_9}. \tag{3.51}
$$

This yields the following system of ODEs giving the dynamics of the high-dimensional GRN, with state variables $x \in \mathbb{R}^9$:

$$
\begin{aligned}
\dot{x}_1 &= \frac{0.2x_1^2 + 0.5 + u_A}{10m + 0.1x_1^2 + 0.5(\gamma_{92}^{-1}x_9)^2 + 0.5x_3^2} - 0.1x_1 \\
\dot{x}_2 &= \frac{0.1x_2^2 + 1 + u_O}{1m + 0.1x_2^2 + 0.5(\frac{x_1 + \gamma_{41}^{-1}x_4}{2})^2 + 0.1x_3^2} - 0.1x_2 \\
\dot{x}_3 &= \frac{0.1x_3^2 + 1 + u_C}{1m + 0.1x_3^2 + 0.5(\gamma_{41}^{-1}x_4)^2 + 0.1(\gamma_{82}^{-1}x_8)^2} - 0.1x_3 \\
\dot{x}_4 &= x_1 + x_4 + x_5 - K_4 x_4 \\
\dot{x}_5 &= x_1 + x_4 + x_5 - K_5 x_5 \\
\dot{x}_6 &= x_4 + x_5 - K_6 x_6 \\
\dot{x}_7 &= x_2 - K_7 x_7 \\
\dot{x}_8 &= x_2 - K_8 x_8 \\
\dot{x}_9 &= x_2 - K_9 x_9
\end{aligned} \tag{3.52}
$$

with γ_{ij} as determined in (3.51), and $K_j, j \in \{4, \ldots, 9\}$ as the free parameters to be chosen.

For the remainder, we choose $K_4 = K_5 = 3, K_6 = K_7 = K_8 = K_9 = 1$. Furthermore, the default values for the other parameters are $m = 1, u_A = u_O = u_C = 0$, corresponding to low levels of the stem cell maintenance factor, and no application of cell type specific stimuli.

Multistability and bifurcation analysis The steady states of the low-dimensional system were obtained by numerically solving the system of equations $\dot{z} = 0$. They are listed, along with their eigenvalues, in Table 3.2. The first three steady states are, according to their eigenvalues, stable, and correspond to the adipogenic, osteogenic, and chondrogenic cell type, as defined by their gene expression levels above. The steady states of the high-dimensional system, along with their eigenvalues, are depicted in Table 3.3. From there it can be seen that the high-dimensional system indeed shares the same multistability properties as the low-dimensional system. While the steady states of the two systems are equal in their first $n = 3$ components, the eigenvalues may slightly differ in their exact values but share the same signs and are, for the chosen parameters, close to the eigenvalues of the low-dimensional system.

In order to investigate the changes in steady states upon changes in the parameters, we conducted bifurcation analysis of the low-dimensional system (using the software package CL_MatCont, Dhooge et al. (2003)). The bifurcation diagrams in Figure 3.4 depict the effect of an osteogenic stimulus u_O: As long as no osteogenic stimulus is applied ($u_O = 0$), there are three stable steady states, and two unstable steady states, as also predicted by the steady state analysis above. For a sufficiently high osteogenic stimulus ($u_O > u_O^{crit} \approx 4.2$), only the osteogenic cell type remains as a stable steady state. Thus, the system will converge towards the osteogenic cell type, which is maintained when the stimulus is withdrawn. Analogous results hold for the effects of an adipogenic stimulus, u_A, and a chondrogenic stimulus, u_C (results not shown).

These results are in accordance with the system properties (R1)-(R3) that were required to be captured by the model. In addition, the effect of a stem cell maintenance factor m can be investigated similarly. Bifurcation analysis reveals that, for a low stem cell maintenance factor $m = 1$, there are the three stable cell types as mentioned, and seen in Figure 3.4. For high levels of the stem cell maintenance factor $m > m^{crit} \approx 4.5$, however, only one stable steady state remains with low levels in all three cell type specific genes, corresponding to a stem cell state (results not shown).

The bifurcation analysis conducted on the low-dimensional system exemplifies how the dependency of multistability properties on parameters can be investigated in the core motif model. As a result of the construction procedure, ensuring multistability equivalence of the high-dimensional system, we know that also the high-dimensional system will have the same multistability properties for the discussed high values of the parameters u_O, m, and also u_A and u_C. It has to be emphasized that, at a saddle-node or pitchfork bifurcation point, there will be an imaginary eigenvalue, thus the multistability equivalence will not hold at exactly this point. However, for parameter values that are not in the critical region, the multistability equivalence is fulfilled, and we can conclude about the multistability properties of the high-dimensional system from the multistability properties of the low-dimensional system. In this way, we have circumvented the need for an exhaustive bifurcation analysis of the high-dimensional system.

Dynamical simulations Exemplary simulations of dynamics are shown in Figure 3.5, for initial values $z(0) = [2.2, 5, 2]^T$, and $x(0) = [2.2, 5, 2, 1, 1, 1, 0, 1, 1]^T$, respectively. As can be seen, the dynamics of the master genes in the high-dimensional system are quite similar to the dynamics in the low-dimensional system, as has been observed also in simulations for other initial values. This similarity between the dynamics could already be expected from the high similarity in the eigenvalues of the systems, recalling Tables 3.2 and 3.3. For choosing very

(a)	$z^{*,1}$	$z^{*,2}$	$z^{*,3}$	$z^{*,4}$	$z^{*,5}$
$z_1^{*(r)}$	12.00	0.08	0.08	7.67	0.12
$z_2^{*(r)}$	0.14	9.90	1.01	0.33	5.67
$z_3^{*(r)}$	0.14	1.01	9.90	0.33	5.67

(b)	λ_f^1	λ_f^2	λ_f^3	λ_f^4	λ_f^5
$\lambda_1^{(r)}$	-0.02	-0.11	-0.10	+0.02	-0.10
$\lambda_2^{(r)}$	-0.10	-0.10	-0.11	-0.10	-0.12
$\lambda_3^{(r)}$	-0.10	-0.07	-0.07	-0.10	+0.05

Table 3.2: Steady states (a) and their eigenvalues (b) of the low-dimensional system.

(a)	$x^{*,1}$	$x^{*,2}$	$x^{*,3}$	$x^{*,4}$	$x^{*,5}$
$x_1^{*(r)}$	12.00	0.08	0.08	7.67	0.12
$x_2^{*(r)}$	0.14	9.90	1.01	0.33	5.67
$x_3^{*(r)}$	0.14	1.01	9.90	0.33	5.67
$x_4^{*(r)}$	12.00	0.08	0.08	7.67	0.12
$x_5^{*(r)}$	12.00	0.08	0.08	7.67	0.12
$x_6^{*(r)}$	24.00	0.17	0.17	15.35	0.24
$x_7^{*(r)}$	0.14	9.90	1.01	0.33	5.67
$x_8^{*(r)}$	0.14	9.90	1.01	0.33	5.67
$x_9^{*(r)}$	0.14	9.90	1.01	0.33	5.67

(b)	λ_F^1	λ_F^2	λ_F^3	λ_F^4	λ_F^5
$\lambda_1^{(r)}$	-0.02	-0.11	-0.10	+0.02	-0.10
$\lambda_2^{(r)}$	-0.10	-0.10	-0.11	-0.10	-0.13
$\lambda_3^{(r)}$	-0.10	-0.07	-0.07	-0.10	+0.05
$\lambda_4^{(r)}$	-1.00	-1.00	-1.00	-1.00	-1.00
$\lambda_5^{(r)}$	-1.00	-1.00	-1.00	-1.00	-1.00
$\lambda_6^{(r)}$	-1.00	-1.00	-1.00	-1.00	-1.00
$\lambda_7^{(r)}$	-1.00	-1.00	-1.00	-1.01	-1.00
$\lambda_8^{(r)}$	-1.00	-1.00	-1.00	-1.00	-1.00
$\lambda_9^{(r)}$	-3.00	-3.00	-3.00	-3.00	-3.00

Table 3.3: Steady states (a) and their eigenvalues (b) of the high-dimensional system. The steady states of the high-dimensional system are equal in their first $n = 3$ components to the steady states of the low-dimensional system. The eigenvalues slightly differ between both systems in their exact values, but share the same signs.

high values for the degradation rate parameters of the module genes, $K_4, \ldots, K_9$, the dynamics of the master genes approach the dynamics of the low-dimensional system even more (results not shown). For such parameter values it was furthermore shown in a similar example by Jouini (2013) that for each steady state of the high-dimensional system, the Nyquist curve approaches the Nyquist curve of the corresponding steady state of the low-dimensional system. These findings on the dynamics and the Nyquist curves both illustrate the property exploited in the proof of (iii), namely that by letting $K_j \to \infty$, $j \in \{n+1, \ldots, N\}$, the Nyquist curves of the two systems can be made arbitrarily close.

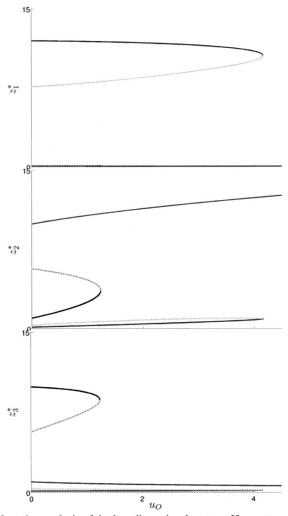

Figure 3.4: Bifurcation analysis of the low-dimensional system: If no osteogenic stimulus is applied ($u_O = 0$), there are three stable steady states (red, black, blue) and two unstable steady states (green dotted, cyan dotted). The three stable steady states correspond to three cell types: An adipogenic cell type (blue; high x_1, low x_2, x_3), an osteogenic cell type (red; high x_2, low x_1, x_3), and a chondrogenic cell type (black; high x_3, low x_1, x_2). Upon increasing the osteogenic stimulus u_O, first the chondrogenic cell type vanishes, then also the adipogenic cell type vanishes, each by a saddle-node bifurcation. Thus, for high values of osteogenic stimulus $u_O > 4.2$, only the osteogenic cell type remains as a stable steady state, and thus the system will converge towards the osteogenic cell type. If the osteogenic stimulus is withdrawn, this stable cell type is maintained.

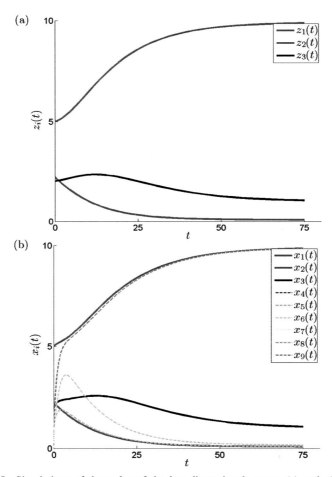

Figure 3.5: Simulations of dynamics of the low-dimensional system (a) and of the high-dimensional system (b). Parameters and initial values as in the text. The dynamics of the master genes in the high-dimensional system are very similar to the dynamics in the low-dimensional system.

3.4   Summary and discussion

In this chapter, we presented a construction method for multistability-equivalent GRNs of different dimensionality. Given the dynamics of a low-dimensional GRN which represents a core motif model, and the interaction structure of a high-dimensional GRN which represents a more detailed description of the system, the proposed method serves to construct for the high-dimensional GRN a dynamical model with the same multistability properties, in terms of steady states and their stability. With this, we provide a solution to the posed Problem 3.1. Our method can be applied to a broad class of GRN systems and provides a novel approach to translate results from core motif models to more realistic GRN models.

The construction procedure is based on the idea that additionally introduced dynamics should be chosen such that their steady state gains are exactly compensated when the system is at steady state. For the presented construction rules as well as for the arguments exploited in the proof, classical tools from systems and control theory were employed, such as the characterization of dynamical systems via transfer functions and their analysis via the Nyquist criterion. Thereby, our contribution provides an example how systems and control theory can serve to develop new approaches in mathematical modeling, and thus enrich modeling-based disciplines such as systems biology.

By studying an example of a GRN in mesenchymal stem cell differentiation, we demonstrated the potential and value of our method. A low-dimensional core motif GRN was employed to taylor a model to reproducing the properties as observed in the biological system, and to investigate the effects of parameter changes as induced for example by differentiation stimuli. A high-dimensional GRN model was then derived via the proposed method and can be fit to the gene expression readout of differentiation experiments. With this, we showed how the results obtained from low-dimensional core motif models can be transferred to more realistic and detailed high-dimensional models of GRNs.

As the construction method proposed here represents sufficient, but not necessary conditions for multistability equivalence, alternative construction methods may be developed. These may well exploit additional degrees of freedom that were not pursued in this work. For example, the linear interaction functions for module genes as in (3.12) may be generalized by weighted sums $\sum_{j=1,\dots,N} \alpha_{ij} x_j$, with more general weights $\alpha_{ij} \in \mathbb{R}$ instead of the coefficients $S_A|_{i,j} \in \{-1, 0, +1\}$. Even more general, the interaction functions may as well be allowed to be nonlinear functions. Such generalizations of the method would provide even more flexibility, however, the validity of such an adjusted construction procedure had to be proven first.

Another open question is whether and under which circumstances equivalences may also be reasoned for bifurcations and vector fields. For example, related questions regarding saddle-node bifurcations in systems of different dimensionality but in the context of model reduction, have been considered by Chiang and Fekih-Ahmed (1993). The example investigated here provides evidence that bifurcations and dynamics may, at least to some extent, have similar properties between multistability-equivalent systems. The exploration of such equivalences with rigorous means for the here considered class of GRN systems provides tempting topics for future work.

Chapter 4

A general model class for proliferating multi-cell type populations in labeling experiments

In this chapter, we develop a general model class for proliferating multi-cell type populations in labeling experiments. By multi-cell type populations we mean cell populations comprising several cell types that differ in their proliferation properties. As proliferation properties are commonly studied by so called labeling experiments, we derive a model that at the same time covers cell types, division numbers, and the label distribution. Section 4.1 provides an overview on the background of multi-cell type populations, proliferation studies, and available models, after which we formulate the problem to be addressed in this chapter. In Section 4.2, we introduce the model class, and present approaches for the analysis and solution. The model class is set in relationship to existing models, which it contains as subclasses or special cases, and to gene regulatory models, in Section 4.3. The chapter is concluded by a short summary and discussion in Section 4.4.

This chapter is partly on the basis of Schittler et al. (2011, 2012, 2013a).

4.1 Background and problem formulation

4.1.1 Multi-cell type populations

Organisms consist of multiple cell types that not only differ in their specific functions, but also in proliferative properties such as division and death rates or average life spans. The typical life span of a cell in humans ranges from few days for blood cells (Shemin and Rittenberg, 1946; Kline and Cliffton, 1952) to decades for bone cells (Parfitt, 1994). As the average life span in turn is determined by division and death rates, this means that also these proliferation rates vary over orders of magnitudes. Not only within an organism, but also within more selected cell populations as studied for example in stem cell research, tissue remodeling, immunology, or cancer therapy, several subpopulations may be present that differ remarkably with respect to their functional properties but also proliferative activity.

For example, T cell populations studied in immunology consist of short-lived effector cells and long-lived memory cells (Hellerstein et al., 2003), whereas upon infection the proportions of these subpopulations change (Debacq et al., 2002; Hellerstein et al., 2003). Also in T cell populations, CD4+ cells were found to proliferate slower than CD8+ cells (De Boer et al., 2003a), and the neglection of such kinetically differing subpopulations in mathematical models was shown to yield wrong estimates of proliferation rates (Asquith et al., 2002). Even if cells are sorted via a specific marker, as commonly done, with the purpose that theoretically all remaining cells should be of the same cell type, practically such markers are not always unique and thus the population may still comprise more than one cell type (De Boer et al., 2012).

Other important examples arise in the field of stem cell research, where especially hematopoietic stem cells have been studied extensively during the last decade: Stem cells may divide asymmetrically (Morrison and Kimble, 2006), meaning that the two daughter cells are of different cell types, such as a stem and a differentiated cell. Furthermore, hematopoietic stem cells were found to switch repeatedly between a quiescent and a proliferating state (Wilson et al., 2008). Consequently, mathematical models considering only one cell type failed to reproduce data of hematopoietic stem cell proliferation assays (Glauche et al., 2009; Kiel et al., 2007; Wilson et al., 2008). The observation that models considering two cell types were able to explain proliferation data better than models with one cell type was made in studies of T cells (Ganusov and De Boer, 2013) as well as of hematopoietic stem cells (Glauche et al., 2009). Besides T cell and stem cell populations, also models for tumors with two cell types were introduced in order to explain the observed population dynamics (Gyllenberg and Webb, 1990).

Other authors noted that cell type transitions frequently occur concomitantly with cell proliferation, constituting important processes in stem cell-induced tissue remodeling (Buske et al., 2011), with diversity ensuring the fitness of a cell population (Balázsi et al., 2011). Not only proliferation rates were shown to depend on division numbers (Hayflick, 1965, 1979), but also, for example, T helper cell differentiation (Bird et al., 1998). Further studies, such as by Buske et al. (2011), highlight the importance of considering cell type transitions and cell proliferation as highly related processes. Overall, this suggests that multiple cell types with clearly distinct proliferation properties may be present in many cell populations, and therefore need to be considered explicitly in mathematical models of such proliferating populations.

4.1.2 Studies on proliferating cell populations

The composition as well as the size of cell populations results to a vast extent from proliferation: Cell division increases the number of the respective types of cells, whereas it is diminished by cell death. Therefore cell proliferation has been a major research subject for several decades. Early quantitative studies of cell proliferation were simply based on cell counts, to which for example exponential growth models were fitted (Zwietering et al., 1990).

More sophisticated studies have been enabled by the establishment of so-called labeling techniques. In these techniques, molecules that under normal conditions are not contained in cells are brought into the cells where they remain, depending on the specific molecule, in the DNA or the cytoplasm. Molecules commonly used as labeling markers are Bromodeoxyuridine (BrdU), Carboxyfluorescein succinimidyl ester (CFSE), deuterated glucose, or deuterated water. These molecules have two important characteristics: Firstly, they are not present in cells under natural conditions. Secondly, they can be detected for example by flow cytometry (as for BrdU and CFSE) or mass spectrometry (as for deuterium). The essential principle of these labeling techniques is that upon cell division, the concentration of the marker molecule is altered, and thus the measured concentration allows to conclude about the number of undergone cell divisions. The two most common labeling techniques, CFSE and BrdU, are discussed in more detail in Chapters 5 and 6.

Recently, experimental techniques have been introduced that allow to trace both cell lineage as well as proliferation dynamics simultaneously (Blanpain and Simons, 2013, and references therein). Such cell lineage markers may allow to trace the offspring of a precursor cell, but the descending cells do not necessarily inherit the same cell type. However, most proliferation studies continue to rely on the established markers such as BrdU and CFSE, and do not simulateously resolve proliferation and cell type structures of a population. The more it is of interest to develop powerful mathematical models for proliferation studies that allow to quantitatively reconstruct knowledge about cell type structure, such as the number of distinct cell types, and their respective proliferation rates.

4.1.3   Available models for multi-cell type populations

For many decades, mathematical models have been used to describe proliferation dynamics (Smith and Martin, 1973; von Foerster, 1959; Zwietering et al., 1990), and to quantify important proliferation parameters such as cell cycle times, growth rates, or division and death rates. Already in the 90s, it was noted that distinguishing between quiescent and actively proliferating subpopulations of cells might be crucial for accurate models and estimates of proliferation parameters, and this encroached on according growth models for example of lymphocyte populations (De Boer and Perelson, 1995) or tumors (Gyllenberg and Webb, 1990).

Following this, models with multiple cell types were used especially in the context of immunologically relevant cell populations, such as to distinguish proliferative activity between T, B and NK cells, CD4+ versus CD8+ cells, naive versus memory cells, or generally resting versus activated cells (Bonhoeffer et al., 2000; De Boer and Perelson, 1995; De Boer and Noest, 1998; De Boer et al., 2003a,b,c; Ribeiro et al., 2002). The majority of these models was tailored to BrdU labeling, some considered telomere lengths (De Boer and Noest, 1998) or deuterium labeling (Ribeiro et al., 2002). Although a central aspect of these models was to investigate proliferation, thus division numbers or/and label concentration would be a natural property to consider, usually neither of these properties was accounted for in these cell type-structured models. Only some models pointed towards this direction by modeling a shortening index of telomeres (De Boer and Noest, 1998), a limited number of division rounds (Grossman et al., 1999; Jones and Perelson, 2005), or the number of labeled DNA strands (Ribeiro et al., 2002). Few models have been introduced since then which incorporate both, multiple cell types as well as principally unlimited division numbers (Glauche et al., 2009; Ganusov and De Boer, 2013).

On the other hand, a multitude of models has been established, in particular tailored to CFSE labeling, that account for division numbers (Deenick et al., 2003; Ganusov et al., 2005; Gett and Hodgkin, 2000; Lee and Perelson, 2008; León et al., 2004; Luzyanina et al., 2007a; Revy et al., 2001; Yates et al., 2007), or label concentration (Luzyanina et al., 2007b, 2009; Banks et al., 2010, 2011), or even both (Schittler et al., 2011; Hasenauer et al., 2012a,b; Hasenauer, 2012; Thompson, 2011). However, none of these models considers cell types, despite their importance as outlined above. Other models explicitly considering cell types have been developed for stem cells and their progressing maturation stages (Marciniak-Czochra et al., 2009; Stiehl and Marciniak-Czochra, 2011), but they in turn do not provide any connection to labeling data.

To the best of the author's knowledge, so far no models were available that consider all three aspects of cell types, division numbers, and label concentration. In addition, it would be beneficial to account for subsequent time intervals and track division numbers therein separately, as it is for example necessary for the uplabeling and delabeling phase in BrdU labeling experiments. Summarizing, a broad variety of models is available, but up to date there is no unifying framework from which specific models for common cell population studies could be deduced, and that at the same time allows to develop solution and analysis tools for a whole model class, independently of the specific problem at hand. Having such a general model class would also support to transfer analytical results and dynamical behavior between related models.

4.1.4 Problem formulation

As pointed out, proliferating multi-cell type populations are the subject of highly relevant studies in multiple fields of biology, and several experimental techniques have been established to study such cell populations. However, there is a lack of suitable mathematical models that consider all of the essential aspects. We therefore formulate the following problem to be addressed in this chapter.

Problem 4.1 *(**Modeling of labeled proliferating multi-cell type populations**) Given the cell type transition properties and the proliferation properties of a cell population, as well as the properties of a label, describe the evolution over time of the labeled, proliferating, multi-cell type population, while accounting for label dynamics, as well as cell type- and division-dependent parameters.*

The present chapter focuses on the general development and analysis of such a model class, whereas models based on this and specified for labeling techniques will be advanced in the two chapters thereafter.

4.2 Development and analysis of a general model class

Dedicated to the above formulated problem, we now develop and analyze a general model class for proliferating multi-cell type populations in labeling experiments.

4.2.1 Development of the model class

The processes in labeled proliferating multi-cell type populations, which we aim to consider, are illustrated in Figure 4.1. During time intervals where label is administered, cells take up label from the environment upon division and thus accumulate more and more label with increasing division number. Conversely, during time intervals where label is withdrawn from the environment, the label content is segregated to daughter cells upon division and thus is reduced with increasing division number. There may be even more than two time intervals with distinct experimental labeling conditions, for example if the label is administered in repeated pulses. To be as general as possible, we will consider a sequence of K time intervals

$(T_{k-1}, T_k], k = 1, \ldots, K$, with $T_0 = 0$, each with its own constant labeling conditions. In addition to cell division, cells may change their cell type, either upon division or as a self-contained process.

To account for these processes, the general model class that we will develop in the following should describe cells in a population with respect to

- the cell type that a cell belongs to j (discrete),

- the number of divisions i_k (discrete) that a cell has undergone in the time interval $(T_{k-1}, T_k]$, and

- the label concentration x (continuous).

For K time intervals $(T_{k-1}, T_k], k = 1, \ldots, K$, the division numbers constitute a division number vector or division sequence, which we will denote by $\mathbf{i} := [i_1, \ldots, i_K]$. While the number of cell divisions i_k is theoretically unlimited, $i_k \in \mathbb{N}_0$, the number of cell types j is finite, $j \in \{1, \ldots, J\}$. To characterize the model class by the processes considered therein, it will be termed as the class of division-sequence-, cell type-, and label-structured population models, for short DsCL models[1].

To describe the considered properties, the state of an individual cell is defined by $(x, \mathbf{i}, j)$. As the presented model describes the population statistics of the cells contained in the population, the model's state variables are the number densities $n(x, \mathbf{i}, j|t)$ of observing a cell with label concentration x, division numbers $\mathbf{i}$, and cell type j, at time t. This number density changes due to label degradation, as well as the fluxes between distinct subpopulations defined by cell types and division numbers, as follows: When a cell of division number $\mathbf{i}$ divides during the k-th time interval $(T_{k-1}, T_k]$, it basically disappears and two cells of division number $\mathbf{i} + \mathbf{e}_k^T$ appear. This means that the two newborn cells have the same division number vector as their mother cell except that the k-th entry from the mother cell is increased by one, $i_k + 1$. Furthermore, a cell of type j_1 can transit into another cell type j_2. Cell type transitions can in principal occur without or together with a cell division. At the beginning of the experiment $(T_0 = 0)$, cells can not have divided yet, thus there are only undivided cells which constitute an initial number density $n_{\text{ini},j}(x) \in \mathbb{R}_+$.

To capture the label uptake and inheritance in a general way, these properties will be represented by distributions. The label uptake usually depends on the label provided in the environment, which in turn may change with time, thus will be denoted by the probability density $p_{\text{lab}}(x|t)$. As a simple example, if the environment contains no label at some time t, there will be no label uptake, hence $p_{\text{lab}}(x|t) = \delta(x)$. The inheritance of label arises from cell division: If cells of division number $\mathbf{i} - \mathbf{e}_k^T$ divide during the time interval $(T_{k-1}, T_k]$, cells of division number $\mathbf{i}$ occur which will inherit a certain amount of label from their mother cell. The label inheritance may possibly depend on the number of divisions and the cell type of the mother cell, as well as on time, and will thus be denoted by $p_{\text{inhe}}(x|\mathbf{i} - \mathbf{e}_k^T, \mu, t)$. For example, if the label gets exactly halved to each daughter cell, then the number density of potential influx, that is the label inheritance times the number of potential mother cells, would be $p_{\text{inhe}}(x|\mathbf{i} - \mathbf{e}_k^T, \mu, t) \int_{\mathbb{R}_+} n(\zeta, \mathbf{i} - \mathbf{e}_k^T, j|t) d\zeta = 2n(2x, \mathbf{i} - \mathbf{e}_k^T, \mu|t)$. Since dividing cells may furthermore take up label upon division, the influx altogether depends on the convolution of label

---

[1]The term "division sequence" is used to denote the vector of division numbers during subsequent time intervals.

uptake and label inheritance, $\int_{\mathbb{R}_+} p_{\text{lab}}(x - \chi|t) p_{\text{inhe}}(\chi|\mathbf{i} - \mathbf{e}_k^T, \mu, t) \int_{\mathbb{R}_+} n(\zeta, \mathbf{i} - \mathbf{e}_k^T, \mu|t) d\zeta d\chi$, and furthermore on division- and cell type transition rates. Although for some labeling techniques (as CFSE) the label inheritance may directly depend on the label concentration of the mother cells, this does not hold for others (such as DNA labeling techniques, as the label inheritance therein further depends on chromosome segregation).

The DsCL model should capture the following processes, with parameters dependent on the concentration x, on time t, on cell type j, or/and on division number $\mathbf{i}$, as indicated:

- label decay, with decay rate $\nu(x, t) \in \mathbb{R}$,

- cell division, with division rates $\alpha_{\mathbf{i}}^j(t) \in \mathbb{R}_+$,

- cell death, with death rates $\beta_{\mathbf{i}}^j(t) \in \mathbb{R}_+$,

- cell type transition (without cell division) from cell type j_1 to cell type j_2, with transition probabilities $\delta_{\mathbf{i}}^{j_2 j_1}(t) \in [0, 1]$, with $\sum_{j_2=1}^{J} \delta_{\mathbf{i}}^{j_2 j_1}(t) = 1 \; \forall \mathbf{i}, j_1, t$,

- cell type transition upon division from cell type j_1 to cell type j_2, with transition-upon-division probabilities $w_{\mathbf{i}}^{j_2 j_1}(t) \in [0, 1]$, with $\sum_{j_2=1}^{J} w_{\mathbf{i}}^{j_2 j_1}(t) = 1 \; \forall \mathbf{i}, j_1, t$,

- label uptake upon division, with an effective label uptake distribution $p_{\text{lab}}(x|t)$,

- label inheritance from mother cells, with a label inheritance distribution $p_{\text{inhe}}(\chi|\mathbf{i} - \mathbf{e}_k^T, \mu, t)$ and the number of potentially dividing cells, $\int_{\mathbb{R}_+} n(\zeta, \mathbf{i} - \mathbf{e}_k^T, \mu|t) d\zeta$.

Taking together the processes outlined above, the DsCL model represents the dynamics of the number density $n(x, \mathbf{i}, j|t)$ under these processes and is given by a system of PDEs, $\forall \mathbf{i} \in \mathbb{N}_0^K, \forall j \in \{1, \ldots, J\}$:

$t \in (T_{k-1}, T_k]:$

$$\frac{\partial n(x, \mathbf{i}, j|t)}{\partial t} + \frac{\partial(\nu(x, t) n(x, \mathbf{i}, j|t))}{\partial x} =$$

$$- \sum_{\mu=1}^{J} \delta_{\mathbf{i}}^{\mu j}(t) n(x, \mathbf{i}, j|t) + \sum_{\mu=1}^{J} \delta_{\mathbf{i}}^{j\mu}(t) n(x, \mathbf{i}, \mu|t)$$

$$- \beta_{\mathbf{i}}^j(t) n(x, \mathbf{i}, j|t) - \alpha_{\mathbf{i}}^j(t) n(x, \mathbf{i}, j|t)$$

$$+ \begin{cases} 0 & , i_k = 0 \\ 2 \sum_{\mu=1}^{J} \left(\alpha_{\mathbf{i}-\mathbf{e}_k^T}^{\mu}(t) w_{\mathbf{i}-\mathbf{e}_k^T}^{j\mu}(t) \int_{\mathbb{R}_+} p_{\text{lab}}(x - \chi|t) \right. \\ \qquad \left. \cdot p_{\text{inhe}}(\chi|\mathbf{i} - \mathbf{e}_k^T, \mu, t) \int_{\mathbb{R}_+} n(\zeta, \mathbf{i} - \mathbf{e}_k^T, \mu|t) \, d\zeta \, d\chi \right) & , i_k \geq 1 \end{cases}$$

$$\tag{4.1}$$

with initial conditions

$$\mathbf{i} = \mathbf{0}^T : n(x, \mathbf{0}^T, j|0) = n_{\text{ini}, j}(x), \; \forall \mathbf{i} \neq \mathbf{0}^T : n(x, \mathbf{i}, j|0) = 0. \tag{4.2}$$

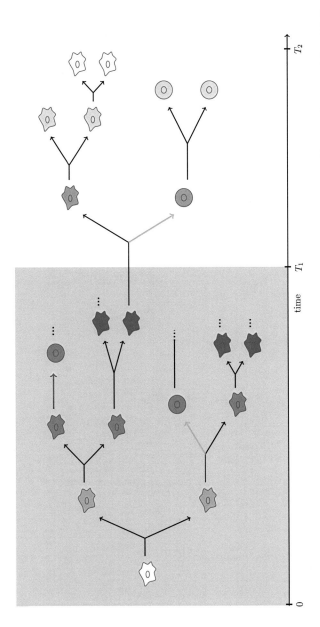

Figure 4.1: Proliferation of labeled multi-cell type populations, exemplarily illustrated for a population with two cell types and for a labeling experiment that comprises two time intervals with different labeling conditions. The first time interval provides label in the environment, denoted by blue background color, whereas in the second time interval the label is withdrawn from the environment, denoted by white background. The two cell types are indicated by star-shaped cells and round cells, respectively. Cell type transition as a self-contained process is denoted by a red arrow, whereas cell type transitions upon division are denoted by green arrows. During the first time interval with label uptake, the cells accumulate label with increasing division number. After the first time interval, only the proliferation of one cell is pursued as a starting point for the second time interval, to facilitate a clear representation. During the second time interval in label-free environment, cells have less label with increasing division number. The label content of cells is in this way correlated with the number of undergone divisions, and is used in proliferation experiments to conclude about proliferation parameters.

The dynamics of $n(x, \mathbf{i}, j|t)$ in this model are then determined by fluxes, as explained in the following:

- $\partial \left(\nu(x,t) n(x, \mathbf{i}, j|t) \right) / \partial x$: decay of label x with decay rate $\nu(x,t)$,

- $\pm \delta_{\mathbf{i}}^{j_2 j_1}(t) n(x, \mathbf{i}, j_1|t)$: transitions from cell type j_1 to j_2 (independent of cell division),

- $-\beta_{\mathbf{i}}^j(t) n(x, \mathbf{i}, j|t)$ and $-\alpha_{\mathbf{i}}^j(t) n(x, \mathbf{i}, j|t)$: cell death and cell division in the subpopulation $(\mathbf{i}, j)$,

- $p_{\text{inhe}}(\chi | \mathbf{i} - \mathbf{e}_k^T, \mu, t) \int_{\mathbb{R}_+} n(\zeta, \mathbf{i} - \mathbf{e}_k^T, \mu|t) \, d\zeta$: cells from the subpopulation $(\mathbf{i} - \mathbf{e}_k^T, \mu)$ divide and give rise to cells with inherited label concentration χ,

- $\int_{\mathbb{R}_+} p_{\text{lab}}(x - \chi|t) \dots d\chi$: additional label uptake upon division, and

- $2 \sum_{\mu=1}^J \alpha_{\mathbf{i} - \mathbf{e}_k^T}^\mu(t) w_{\mathbf{i} - \mathbf{e}_k^T}^{j\mu}(t) \dots$: the appearance of two newborn daughter cells of cell type j, with division rate $\alpha_{\mathbf{i} - \mathbf{e}_k^T}^\mu(t)$ and transition-upon-division probability $w_{\mathbf{i} - \mathbf{e}_k^T}^{j\mu}(t)$ from cell type μ.

We briefly review the information that is provided in the DsCL model:

- The *number of cells in each subpopulation*, defined by division number and cell type, $(\mathbf{i}, j)$, is

$$N(\mathbf{i}, j|t) = \int_{\mathbb{R}_+} n(x, \mathbf{i}, j|t) dx. \tag{4.3}$$

From that, also the number of cells with certain division number $\mathbf{i}$ (regardless of cell type) can be obtained by summing over all cell types,

$$\underline{N}(\mathbf{i}|t) = \sum_{j=1}^J N(\mathbf{i}, j|t), \tag{4.4}$$

and the number of cells of certain cell type j (regardless of division number) can similarly be obtained by summing over all division numbers,

$$\overline{N}(j|t) = \sum_{i_1=0}^\infty \dots \sum_{i_K=0}^\infty N(\mathbf{i}, j|t). \tag{4.5}$$

- The *accumulated label distribution* of cells with certain division number $\mathbf{i}$ (regardless of cell type),

$$\underline{n}(x, \mathbf{i}|t) = \sum_{j=1}^J n(x, \mathbf{i}, j|t), \tag{4.6}$$

and the accumulated label distribution of cells of certain cell type j (regardless of division number),

$$\overline{n}(x, j|t) = \sum_{i_1=0}^\infty \dots \sum_{i_K=0}^\infty n(x, \mathbf{i}, j|t), \tag{4.7}$$

may be of interest.

- The *label density in each subpopulation*, that is the probability density over x, given that a cell belongs to subpopulation $(\mathbf{i}, j)$ at time t, is for all $(\mathbf{i}, j, t) \in \mathbb{N}_0^K \times \{1, \ldots, J\} \times \mathbb{R}_+$ for which $N(\mathbf{i}, j|t) > 0$, given by

$$p(x|\mathbf{i}, j, t) = \frac{n(x, \mathbf{i}, j|t)}{N(\mathbf{i}, j|t)}. \tag{4.8}$$

Similarly, the label density of cells with certain division number $\mathbf{i}$ (regardless of cell type) and of cells of certain cell type j (regardless of division number) can be obtained via the respective accumulated label distribution and number of cells.

- The *overall label distribution* in the population is given by

$$m(x|t) = \sum_{j=1}^{J} \sum_{i_1=0}^{\infty} \ldots \sum_{i_K=0}^{\infty} n(x, \mathbf{i}, j|t). \tag{4.9}$$

- The *overall number of cells* in the population is given by

$$M(t) = \sum_{j=1}^{J} \sum_{i_1=0}^{\infty} \ldots \sum_{i_K=0}^{\infty} N(\mathbf{i}, j|t) = \int_{\mathbb{R}_+} m(x|t) dx. \tag{4.10}$$

Two important aspects for the connection to data from labeling experiments should be noted: First, data from these experiments is usually marginalized over division numbers $\mathbf{i}$ and cell types j, since common labeling techniques such as CFSE or BrdU do not allow to distinguish uniquely between cell types nor division numbers. Therefore, the overall label distribution $m(x|t)$ is the variable to be matched to data. Second, the number of cells in this modeling framework is defined on a continuous scale, as opposed to discrete numbers of cells, which is feasible if cell numbers are large enough as it is usually the case in common labeling experiments. The further relationship between label concentration, label-induced fluorescence, and actually measured fluorescence has been discussed in detail especially for the marker CFSE elsewhere (Banks et al., 2010; Hasenauer et al., 2012a).

4.2.2 Analysis of the model class

In the previous section, a general model class has been introduced to describe labeled proliferating multi-cell type populations with respect to their cell type, division number, and label concentration. In general, the DsCL model class is represented by a set of coupled PDEs (4.1), for which the solution and simulation would be computationally highly demanding, if not impossible. Fortunately, in many cases the properties of labeling experiments permit to specify a model within this model class that can be solved much more efficiently. We will investigate this subclass of models, along with its treatment, in the following.

Solution approach via decomposition For common labeling techniques, such as CFSE and BrdU which will be covered in the next two chapters, the system of coupled PDEs (4.1) can be decomposed, enabling a much more efficient solution of the model. It is thus worth to elaborate a more detailed analysis of the subclass of DsCL models for which this decomposition is possible, as the vast majority of applications can be expected to fall into this subclass. The advantage of elaborating the analysis on this general model subclass is clearly that solutions and tools can be developed on the methodological side, independent of the specific application, and will be available for modeling problems that will arise from various applications.

Let us thus proceed with the focus on a subclass of DsCL models, for which it can be proven that these models can be decomposed. Before stating this in a theorem, it is important to note that the probability densities $p(x|i, j, t)$ as in (4.8) are always defined except for the trivial solution case, $n(x, i, j|t) = 0$. That is, if for some $(i, j, t) \in \mathbb{N}_0^K \times \{1, \ldots, J\} \times \mathbb{R}_+$ it is $N(i, j|t) = 0$, then the solution for this (i, j, t) is $\forall x : n(x, i, j|t) = 0$. This follows directly from the definition of $N(i, j|t)$, (4.3), and because the number density $n(x, i, j|t)$ is always nonnegative. Therefore, the equality $0 = N(i, j|t) = \int_{\mathbb{R}_+} n(x, i, j|t)dx$ only holds if $\forall x : n(x, i, j|t) = 0$.

Because of this it only remains to search for solutions $n(x, i, j|t)$ at time points t and for (i, j) for which $N(i, j|t) \neq 0$. This in turn guarantees that $p(x|i, j, t)$ are defined as in (4.8), which is important as these probability densities will be exploited to formulate a solution approach based on a decomposition. The following theorem gives a sufficient condition for the decomposability of the DsCL model, and thereby provides a step towards solving the DsCL model as will be seen thereafter.

Theorem 4.2 *If for all $(i, j, t) \in \mathbb{N}_0^K \times \{1, \ldots, J\} \times (T_{k-1}, T_k]$ for which $N(i, j|t) \neq 0$ it can be rewritten*

$$\int_{\mathbb{R}_+} p_{\text{lab}}(x - \chi|t) p_{\text{inhe}}(\chi|i - e_k^T, \mu, t) \, d\chi = p(x|i, t), \tag{4.11}$$

with for all $j : p(x|i, j, t) = p(x|i, t)$, and for all $t \in (T_{k-1}, T_k] : p_{\text{lab}}(x|t) = p_{\text{lab}}^{(k)}(x)$ invariant during this time interval, then the solution of the DsCL model (4.1) is given by $\forall i, j, t$:

$$n(x, i, j|t) = N(i, j|t) p(x|i, t), \tag{4.12}$$

wherein $N(i, j|t)$ is the solution of the system of ODEs $\forall i \in \mathbb{N}_0^K, \forall j \in \{1, \ldots, J\}$:

$$t \in (T_{k-1}, T_k] :$$

$$\begin{aligned} \frac{dN(i, j|t)}{dt} = &-\sum_{\mu=1}^{J} \delta_i^{\mu j}(t) N(i, j|t) + \sum_{\mu=1}^{J} \delta_i^{j\mu}(t) N(i, \mu|t) \\ &- \beta_i^j(t) N(i, j|t) - \alpha_i^j(t) N(i, j|t) \\ &+ \begin{cases} 0 & , i_k = 0 \\ 2 \sum_{\mu=1}^{J} \alpha_{i-e_k^T}^\mu(t) w_{i-e_k^T}^{j\mu}(t) N(i - e_k^T, \mu|t) & , i_k \geq 1 \end{cases} \end{aligned} \tag{4.13}$$

with initial conditions

$$i = 0^T : N(0^T, j|0) = N_{\text{ini},j}, \ \forall i \neq 0^T : N(i, j|0) = 0, \tag{4.14}$$

and $p(x|i, t)$ is the solution of the set of PDEs $\forall i \in \mathbb{N}_0^K$:

$$\frac{\partial p(x|i, t)}{\partial t} + \frac{\partial (\nu(x, t) p(x|i, t))}{\partial x} = 0, \tag{4.15}$$

with initial conditions

$$\mathbf{i} = \mathbf{0}^T : p(x|\mathbf{0}^T, 0) = p_{\text{ini}}(x),$$

$$\forall \mathbf{i} \neq \mathbf{0}^T \,,\ \text{with}\ \ k' := \max\{k | i_k \geq 1\} : p(x|\mathbf{i}, 0) = \int_{\mathbb{R}_+} p_{\text{lab}}^{(k')}(x - \chi) p_{\text{inhe}}(\chi|\mathbf{i} - \mathbf{e}_{k'}^T, \mu, 0) d\chi,$$

$$(4.16)$$

wherein the label uptake during each time interval $p_{\text{lab}}^{(k')}(x - \chi) = p_{\text{lab}}(x - \chi|T_{k'})$ and the label inheritance $p_{\text{inhe}}(\chi|\mathbf{i} - \mathbf{e}_k^T, \mu, 0)$ are determined by the specific labeling technique.

This theorem states that if the combined result of label inheritance from the dividing cells of subpopulation $(\mathbf{i} - \mathbf{e}_k^T, \mu)$ and label uptake at time t equals the label density of the cells in subpopulation $(\mathbf{i}, j)$ at time t during the k-th time interval, independently of the cell types j, μ, j, then the cell population dynamics can be separated from the label dynamics.

Proof The following proof is conducted for $i_k \geq 1$, since for $i_k = 0$ it works analogously but the last term in the ODE is zero. Inserting (4.12) into the left hand side of (4.1) and using the chain rule yields

$$t \in (T_{k-1}, T_k] :$$
$$\frac{\partial(N(\mathbf{i}, j|t)p(x|\mathbf{i}, t))}{\partial t} + \frac{\partial(\nu(x, t)N(\mathbf{i}, j|t)p(x|\mathbf{i}, t))}{\partial x}$$
$$= \frac{dN(\mathbf{i}, j|t)}{dt}p(x|\mathbf{i}, t) + \frac{\partial p(x|\mathbf{i}, t)}{\partial t}N(\mathbf{i}, j|t) + \frac{\partial(\nu(x, t)p(x|\mathbf{i}, t))}{\partial x}N(\mathbf{i}, j|t)$$
$$= \frac{dN(\mathbf{i}, j|t)}{dt}p(x|\mathbf{i}, t) + \left(\frac{\partial p(x|\mathbf{i}, t)}{\partial t} + \frac{\partial(\nu(x, t)p(x|\mathbf{i}, t))}{\partial x}\right)N(\mathbf{i}, j|t)$$

$$(4.17)$$

Inserting (4.12) into the right hand side of (4.1) and substituting the integral therein by (4.11) gives

$$t \in (T_{k-1}, T_k] :$$

$$-\sum_{\mu=1}^{J} \delta_{\mathbf{i}}^{\mu j}(t)N(\mathbf{i}, j|t)p(x|\mathbf{i}, t) + \sum_{\mu=1}^{J} \delta_{\mathbf{i}}^{j\mu}(t)N(\mathbf{i}, \mu|t)p(x|\mathbf{i}, t) - (\beta_{\mathbf{i}}^{j}(t) + \alpha_{\mathbf{i}}^{j}(t))N(\mathbf{i}, j|t)p(x|\mathbf{i}, t)$$

$$+ 2\sum_{\mu=1}^{J} \alpha_{\mathbf{i}-\mathbf{e}_k^T}^{\mu}(t)w_{\mathbf{i}-\mathbf{e}_k^T}^{j\mu}(t)N(\mathbf{i} - \mathbf{e}_k^T, \mu|t)p(x|\mathbf{i}, t)$$

$$= \left(-\sum_{\mu=1}^{J} \delta_{\mathbf{i}}^{\mu j}(t)N(\mathbf{i}, j|t) + \sum_{\mu=1}^{J} \delta_{\mathbf{i}}^{j\mu}(t)N(\mathbf{i}, \mu|t) - (\beta_{\mathbf{i}}^{j}(t) + \alpha_{\mathbf{i}}^{j}(t))N(\mathbf{i}, j|t)\right.$$

$$\left. + 2\sum_{\mu=1}^{J} \alpha_{\mathbf{i}-\mathbf{e}_k^T}^{\mu}(t)w_{\mathbf{i}-\mathbf{e}_k^T}^{j\mu}(t)N(\mathbf{i} - \mathbf{e}_k^T, \mu|t)\right)p(x|\mathbf{i}, t)$$

$$(4.18)$$

Now, set the left hand side (4.17) equal to the right hand side (4.18). Furthermore, let us shift the terms containing $N(\mathbf{i}, j|t)$ to the left side, and the terms containing $p(x|\mathbf{i}, t)$ to the right side,

$t \in (T_{k-1}, T_k]$:

$$\left(\frac{\partial p(x|\mathbf{i}, t)}{\partial t} + \frac{\partial \nu(x, t) p(x|\mathbf{i}, t)}{\partial x} \right) N(\mathbf{i}, j|t)$$

$$= \left(\frac{dN(\mathbf{i}, j|t)}{dt} - \sum_{\mu=1}^{J} \delta_{\mathbf{i}}^{\mu j}(t) N(\mathbf{i}, j|t) + \sum_{\mu=1}^{J} \delta_{\mathbf{i}}^{j\mu}(t) N(\mathbf{i}, \mu|t) \right.$$

$$\left. - (\beta_{\mathbf{i}}^{j}(t) + \alpha_{\mathbf{i}}^{j}(t)) N(\mathbf{i}, j|t) + 2 \sum_{\mu=1}^{J} \alpha_{\mathbf{i} - \mathbf{e}_k^T}^{\mu}(t) w_{\mathbf{i} - \mathbf{e}_k^T}^{j\mu}(t) N(\mathbf{i} - \mathbf{e}_k^T, \mu|t) \right) p(x|\mathbf{i}, t).$$

$$(4.19)$$

From the PDE (4.15), the expression in the brackets on the left hand side is zero. From the ODE (4.13), the expression in the brackets on the right hand side is zero. With this and regardless of $N(\mathbf{i}, j|t)$, $p(x|\mathbf{i}, t)$, the equation (4.19) reduces to $0 = 0$.

For the initial conditions, we obtain by using (4.12) and inserting (4.14) and (4.16):

$$\begin{aligned} \mathbf{i} = \mathbf{0}^T &: n(x, \mathbf{0}^T, j|0) = N_{\text{ini},j} p_{\text{ini}}(x) = n_{\text{ini},j}(x), \\ \forall \mathbf{i} \neq \mathbf{0}^T &: n(x, \mathbf{0}^T, j|0) = 0 \cdot p(x|\mathbf{i}, 0) = 0, \end{aligned}$$

$$(4.20)$$

which equals the initial conditions (4.2) of the model.

With this we have proven that, given that (4.11) holds, with $N(\mathbf{i}, j|t)$ being the solution of (4.13) and $p(x|\mathbf{i}, t)$ the solution of (4.15), (4.12) solves the DsCL model (4.1). □

Importantly, this decomposition (4.12) substitutes the system of coupled PDEs (4.1) by a system of ODEs (4.13) and a set of decoupled PDEs (4.15). The system of ODEs can be solved efficiently, for example by numerical integration, such that it only remains to solve the set of PDEs. As the PDEs give the label dynamics, these will depend on the applied labeling technique from which the equations (4.15) can be further specified. As will be seen, models specified for common labeling techniques not only satisfy the decomposition property of Theorem 4.2, but the resulting PDEs can even be solved analytically. While it might seem intuitive that in many cases the dynamics of $n(x, \mathbf{i}, j|t)$ can be decomposed, it has to be shown mathematically and it is important to note that for the general case (4.1), (4.2), this does not hold true. Examples where this decomposition can not be conducted in general are if the label decay $\nu(x, t)$ is nonlinear in the label concentration x, or if the label uptake would vary within one time interval, that is there exist time points $t_1, t_2 \in (T_{k-1}, T_k]$ for which $p_{\text{lab}}(x - \chi|t_1) \neq p_{\text{lab}}(x - \chi|t_2)$. In such cases, (4.11) may be vulnerated and the dynamics of the label densities, due to a sink/source term, can not be separated from the population dynamics any more.

The next two chapters, Chapter 5 and 6, will present models tailored to the specific labeling techniques of CFSE and BrdU, and it will be demonstrated that these models indeed fulfill (4.11), thus becoming amenable to the solution approach via the decomposition (4.12). The detailed development and analysis of such specified models is the subject of the respective next chapters. For CFSE labeling, the distributions $p(x|\mathbf{i}, t)$ will be obtained as analytical solutions of a linear PDE, see Chapter 5. For BrdU labeling, we will derive the distributions from modeling the process of label uptake into newly synthesized chromosomes and the label inheritance via chromosome segregation in Chapter 6. Thereupon, it remains to solve the ODE part in order to achieve $N(\mathbf{i}, j|t)$.

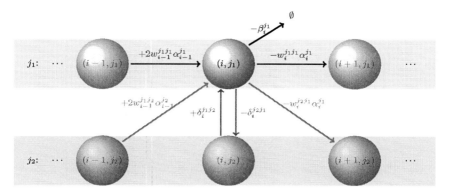

Figure 4.2: Model structure of the DsC model. Illustration of cell types (rows) and division numbers (columns), for one time interval ($K = 1$): The circles represent subpopulations (i, j) of cells with division number i and cell type j, and the arrows denote the fluxes between these subpopulations. Purple: Fluxes due to cell type transitions (without cell division). Green: Fluxes due to cell type transitions upon cell division. Blue: Fluxes due to cell division (without type transition). Black: Outflux due to cell death.

Since the ODE part (4.13) now describes the dynamics of the cell population independent of the specific labeling technique, it is worth analyzing the dynamics of the number of cells $N(\mathbf{i}, j|t)$ in a general context in this chapter, whereas the label dynamics $p(x|\mathbf{i}, t)$ will depend on the labeling technique. The remainder of the current chapter will therefore focus on the analysis of the dynamics of $N(\mathbf{i}, j|t)$, the ODE part of a model decomposed as in (4.12). Since the label dynamics are split from the population dynamics due to the decomposition, the remaining ODE system for the number of cells $N(\mathbf{i}, j|t)$ is structured due to division sequence and cell types. Accordingly, we denote this model part (4.13) as a division-sequence- and cell type-structured population model (DsC model).

The fluxes in and out of the subpopulations are depicted, exemplarily at one subpopulation (i, j_1) within a population with two cell types, j_1 and j_2, in Figure 4.2. The ODE system (4.13), (4.14) is of infinite dimension, since infinitely high division numbers $i_k \in \mathbb{N}_0$ are considered. In principal, there are two possibilities to solve this ODE system: One can either use a truncated ODE system which can then be efficiently solved numerically, or in several cases it is even possible to derive an analytical solution. Both options will be discussed in the following paragraphs.

Truncated ODE system Using a truncated version of the infinitely dimensional system (4.13) means to only consider a finite number, S_k, of divisions for each of the time intervals $k \in \{1, \ldots, K\}$. For simplicity, we consider the same truncation number for each time interval here, $\forall k : S_k = S$, whereas more flexibly one may even choose different truncation values. The dimension of the system's state variable, $N(\mathbf{i}, j|t)$, is proportional to S^K, thus it increases exponentially with the number of time intervals K.

For the truncated version the same ODEs as in (4.13), (4.14) are used but only for $\mathbf{i} \in$

$\{0, \ldots, S-1\}^K$. To distinguish from the exact solution of the full (infinitely dimensional) ODE system, let the overall number of cells of the truncated system be denoted by

$$\hat{M}_S(t) = \sum_{j=1}^{J} \sum_{i_1=0}^{S-1} \cdots \sum_{i_K=0}^{S-1} N(\mathbf{i}, j|t) = \int_{\mathbb{R}_+} \hat{m}_S(x|t) dx, \qquad (4.21)$$

with the overall label distribution resulting from the solution of the truncated system denoted by

$$\hat{m}_S(x|t) = \sum_{j=1}^{J} \sum_{i_1=0}^{S-1} \cdots \sum_{i_K=0}^{S-1} n(x, \mathbf{i}, j|t) = \sum_{j=1}^{J} \sum_{i_1=0}^{S-1} \cdots \sum_{i_K=0}^{S-1} N(\mathbf{i}, j|t) p(x|\mathbf{i}, t). \qquad (4.22)$$

Such a truncated system can be solved via common numerical integration, and its relative truncation error at a certain time point $t \in (T_{k-1}, T_k]$, is upper bounded via

$$\frac{M(t) - \hat{M}_S(t)}{\displaystyle\sum_{j=1}^{J} N_{\text{ini},j}} \leq E_S(t), \qquad (4.23)$$

with the upper bound $E_S(t)$ obtained from solving a bounding system, as outlined in Appendix C. This was shown for one time interval, $K = 1$, and without cell types, $J = 1$, in Hasenauer et al. (2012a), whereas we generalize the proof to arbitrary numbers of time intervals, $K \geq 1$, and of cell types, $J \geq 1$, in Appendix C. Thus, the truncation error can be made arbitrarily small by choosing a sufficiently high S.

Importantly, the obtained error bound not only restricts the error of the solution $\hat{M}_S(t)$ of the truncated ODE system, but also of the corresponding overall label distribution, $\hat{m}_S(t)$, as shown in Appendix C. This is a powerful finding since it states that for decomposable models of the DsCL class, only a finite system of ODEs has to be solved. The numerical solution of a linear ODE system can be handled by common ODE solvers in very low computation time. The upper bound for the truncation error can both be used to estimate the maximum error, as well as to derive the minimally required value of S if it is desired to keep the truncation error below a certain value. As was found in former studies (Hasenauer et al., 2012b) and will also be seen later in the example of Chapter 5, usually a quite moderate value of $S \approx 20$ is sufficient to already ensure a satisfying accuracy of $E_S(t) \leq 10^{-3}$. With this, the class of decomposable DsCL models, into which models for common labeling techniques will fall (see Chapter 5 and 6), becomes attractive also for its efficient solution, as well as its error handling.

Next, we will see how this class of models even allows for analytical solutions in some cases, which opens up new opportunities for investigating the dynamical properties, especially related to parameters.

Analytical solutions   In some cases it may even be possible to solve the full ODE system (4.13) with initial conditions (4.14) analytically. This may bring along the advantage that certain properties of the dynamics, such as growth or decline of the population, become directly visible in the solution equations. Furthermore, parameter subspaces which imply qualitatively distinct system dynamics may be determined from an analytical solution, without the need to conduct simulations and empirical approaches. An example of such a case is demonstrated in the next chapter in Section 5.3. Here, we will discuss some rather general cases for which such an ODE system may even be solved analytically, while these cases may not be exhaustive.

The subclass of models where there are no transitions between cell types,

$$\forall \mathbf{i}, j_1, j_2, t : \delta_{\mathbf{i}}^{j_2 j_1}(t) = 0$$
$$\forall \mathbf{i}, t : w_{\mathbf{i}}^{jj}(t) = 1 \tag{4.24}$$
$$\text{which implies } \forall \mathbf{i}, j_1 \neq j_2, t : w_{\mathbf{i}}^{j_1 j_2}(t) = 0,$$

is closely related to the case of not considering distinct cell types, $J = 1$, since the solution for each cell type is independent of the others. We present analytical solutions for two cases within this subclass.

Proposition 4.3 *If (4.24) holds, and division and death rates are constant within each time interval, $\forall \mathbf{i}, j, t : \alpha_{\mathbf{i}}^j(t) = \alpha_{\mathbf{i}}^j, \beta_{\mathbf{i}}^j(t) = \beta_{\mathbf{i}}^j$, and furthermore $\forall j, \mathbf{i}', \mathbf{i}'', \mathbf{i}' \neq \mathbf{i}'' : \alpha_{\mathbf{i}'}^j + \beta_{\mathbf{i}'}^j \neq \alpha_{\mathbf{i}''}^j + \beta_{\mathbf{i}''}^j$, then the solution of (4.13), (4.14) is given by $\forall t \in (T_{k-1}, T_k]$:*

$$N(\mathbf{i},j|t) = \begin{cases} 2^{(\sum_{\nu=1}^{k} i_\nu)} \left(\prod_{\nu=1}^{k} C_{\mathbf{i},j}^\nu \right) \left(\prod_{\nu=1}^{k-1} D_{\mathbf{i},j}^\nu(T_\nu) \right) D_{\mathbf{i},j}^k(t) N_{\text{ini},j} & , \text{ if } \forall \nu > k : i_\nu = 0 \\ 0 & , \text{ otherwise} \end{cases}$$

$$\text{with } \tilde{\mathbf{i}}^{(\nu,l)} := [i_1, \ldots, i_{\nu-1}, l, 0, \ldots, 0],$$

$$C_{\mathbf{i},j}^\nu := \begin{cases} 1 & , i_\nu = 0 \\ \prod_{l=1}^{i_\nu} \alpha_{\tilde{\mathbf{i}}(\nu,l-1)}^j & , i_\nu \geq 1 \end{cases},$$

$$D_{\mathbf{i},j}^\nu(t) := \begin{cases} e^{-(\alpha_{\tilde{\mathbf{i}}(\nu,0)}^j + \beta_{\tilde{\mathbf{i}}(\nu,0)}^j)t} & , i_\nu = 0 \\ \left(\sum_{l=0}^{i_\nu} \left[\left(\prod_{\substack{j=0 \\ j \neq l}}^{i_\nu} ((\alpha_{\tilde{\mathbf{i}}(\nu,j)}^j + \beta_{\tilde{\mathbf{i}}(\nu,j)}^j) - (\alpha_{\tilde{\mathbf{i}}(\nu,l)}^j + \beta_{\tilde{\mathbf{i}}(\nu,l)}^j)) \right)^{-1} \right. \\ \qquad \left. \cdot e^{-(\alpha_{\tilde{\mathbf{i}}(\nu,l)}^j + \beta_{\tilde{\mathbf{i}}(\nu,l)}^j)t} \right] \right) & , i_\nu \geq 1 \end{cases}.$$

$$\tag{4.25}$$

For the case of one cell type, $J = 1$, and one time interval, $K = 1$, this solution has been presented in Luzyanina et al. (2007a) and the according proof in Hasenauer et al. (2012a). The proof for the general case with arbitrarily many time intervals $K \geq 1$ is provided in Appendix D.

Proposition 4.4 *If (4.24) holds, and division and death rates are independent of time and division number, $\forall \mathbf{i}, j, t : \alpha_{\mathbf{i}}^j(t) = \alpha^j, \beta_{\mathbf{i}}^j(t) = \beta^j$, then the solution of (4.13), (4.14) is given* *by* $\forall t \in (T_{k-1}, T_k]$:

$$
N(\mathbf{i}, j | t) = \begin{cases} \dfrac{(2\alpha^j)^{\left(\Sigma_{l=1}^{k} i_l\right)} \left(\displaystyle\prod_{l=1}^{k-1} (T_l - T_{l-1})^{i_l}\right) (t - T_{k-1})^{i_k}}{\displaystyle\prod_{l=1}^{k} i_l!} \\ \qquad\qquad\qquad\qquad \cdot e^{-(\alpha^j + \beta^j)t} N_{\mathrm{ini},j} \quad, \ if \ \forall \nu > k : i_\nu = 0 \\[2ex] 0 \qquad\qquad\qquad\qquad\qquad\qquad\qquad\quad , \ otherwise \end{cases}
$$

$$(4.26)$$

This result is already known for the special cases of one time interval, $K = 1$, as stated in De Boer et al. (2006); Hasenauer et al. (2012a); Revy et al. (2001); Schittler et al. (2011), and for two time intervals, $K = 2$, shown by Ganusov and De Boer (2013). Here the result extends to the general case $K \geq 1$ and the proof is provided in Appendix E.

Further cases where analytical solutions can be found may arise as long as cell type transitions do not induce circular fluxes, in the meaning that the resulting signal flow graph (Figure 4.2) has no circles. One such case arises in the example studied in the next chapter, in Section 5.3. This will also further highlight the usefulness of analytical solutions of DsCL models.

In contrast, in cases with circular fluxes in the signal flow graph, no analytical solution could be found so far to the best of the author's knowledge. In all cases where no analytical solution is available, or if merely the simulation is of interest instead of analysis, the truncated system with upper-bounded truncation error (4.23), as discussed above, may be used for a numerical solution.

4.3 Relationship to other models and model classes

To promote a better understanding of the DsCL model class, in this section we discuss its relationship to existing population models as well as to gene switch models (treated in Chapter 2 and 3).

4.3.1 Classification of existing population models as subclasses

After we have developed the general model class of division-sequence-, cell type-, and label-structured population models (DsCL models) and analyzed its solution, we will now discuss existing models as subclasses or special cases of DsCL models, and relate them to each other.

A schematic overview of models within the DsCL model class is given in Fig. 4.3. The following discussion of these models makes no claim to be complete. Many applications and adjustments of these models exist, such that further published models may well fall within the model class treated here.

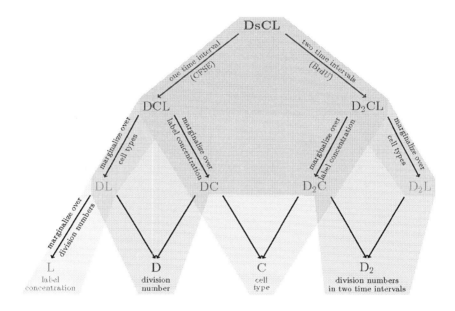

Figure 4.3: Overview of the DsCL model class and existing models that are within the DsCL model class, and relationships to each other. To characterize the model classes, the following abbreviations are used: D - division number, s - subsequent time intervals, C - cell types, L - label. The abbreviations denote the properties that are explicitly considered in the respective model structure. Each arrow indicates a marginalization over one property (that is then lost in the resulting, less structured, model). The D... models have mainly been developed for studying CFSE experiments, whereas the D_2... models mainly arise in modeling of BrdU labeling experiments. Colored shades indicate which properties are considered in the explicit model class: green - label concentration, blue - division number(s), red - cell types. The division-sequence-, cell type-, and label- (DsCL) structured model class, which is presented in this chapter, is the most general one and contains all other models as subclasses.

Division-sequence-, cell type-, and label-structured (DsCL) The DsCL model class has been introduced in the present chapter. It contains all the following models as subclasses and thus provides a general framework to study proliferation in populations that may comprise distinct cell types, while accounting for labeling properties and several time intervals in a proliferation experiment.

For common labeling techniques, such as CFSE and BrdU, it is possible to conduct a decomposition approach for the respective DsCL model. Thus, albeit its generality, it is of only moderate complexity in the specific context where usually only few time intervals have to be considered. Specific DsCL models for the common labeling techniques CFSE and BrdU will be developed in the following chapters and demonstrate how the model can be solved and simulated in a highly efficient way.

The left branch of the model tree in Fig. 4.3 contains specific DsCL models and subclasses that consider one time interval of divisions, whereas the right branch of the model tree collects specific DsCL models comprising two time intervals. In CFSE labeling experiments, usually only one time interval needs to be considered, whereas two time intervals arise typically in BrdU labeling experiments. Models with more than two time intervals arise for example when repeated pulse dosages of label are administered. Although not pursued up to date, also such models would be covered by the general DsCL model class.

DsCL models with one time interval (DCL)   The specification of a DsCL model for CFSE labeling leads to a DCL model, comprising one time interval, as will be outlined in this work in Chapter 5. This model has been presented in Schittler et al. (2012), to the best of the author's knowledge as the first one comprising division numbers, cell types, and label dynamics together. In general, a DCL model is given by the system of PDEs (4.1) with $K = 1$, and the division sequence vector simplifies to a scalar $\mathbf{i} = i$. For the specific DCL model as introduced in Schittler et al. (2012), further labeling specifications for CFSE have been made as can also be found in Chapter 5, and which importantly result in a model that can be decomposed (4.12).

Division- and label-structured (DL)   The DL model has been introduced in the context of CFSE labeling by Schittler et al. (2011); Hasenauer et al. (2012a,b), and Thompson (2011), and also been used for example by Banks et al. (2012). Choosing the general formulation with explicitly distinguishing label inheritance and label uptake as introduced here, it can be written as a system of PDEs, $\forall i \in \mathbb{N}_0, t \in (0, T]$:

$$\frac{\partial \underline{n}(x, i|t)}{\partial t} + \frac{\partial(\nu(x, t)\,\underline{n}(x, i|t))}{\partial x} = -\underline{\beta}_i(t)\underline{n}(x, i|t) - \underline{\alpha}_i(t)\underline{n}(x, i|t)$$

$$+ \begin{cases} 0 & , i = 0 \\ 2\underline{\alpha}_{i-1}^{\mu}(t) \int_{\mathbb{R}_+} p_{\text{lab}}(x - \chi|t)\,\underline{n}_{\text{inhe}}(\chi, i - 1|t)\,d\chi & , i \geq 1 \end{cases}$$

$$(4.27)$$

with initial conditions $\underline{n}(x, 0|0) = \underline{n}_{\text{ini}}(x)$, $\forall i \neq 0 : \underline{n}(x, i|0) = 0$. For the case of CFSE, for which the DL models have been introduced in the mentioned publications, the labeling properties are further specified, similarly as in Chapter 5 for the DCL model. In case of one cell type, $J = 1$, a DCL model equals a DL model, and thus the DL model class is a special case of the more general DCL model class.

Division- and cell type-structured (DC)   Division- and cell type-structured population models do not consider an explicit label concentration, but rather cell numbers of a population structured according to division numbers and cell types. It can thus be derived from a DCL model by marginalizing over the label concentration x, as by (4.3), or if the decomposition (4.12) can be conducted, by separating the label dynamics from the cell population dynamics which then remain in the DC model.

Models of the DC model subclass were introduced, for example, to study telomere loss of populations with naive and memory cells (De Boer and Noest, 1998), cell counts under pathogen growth in effector (CD8+) and target (CD4+) T cells (Jones and Perelson, 2005), and proliferation dynamics in asymmetrically dividing stem cell populations (Schittler et al., 2012). The example studied in the latter will also be analyzed in detail in this work: In the example in Chapter 5 a DC model serves to identify parameter subsets with qualitatively distinct cell population dynamics.

The general form of a DC model is given by a system of ODEs, $\forall i \in \mathbb{N}_0, j \in \{1, \ldots, J\}$:

$$\frac{dN(i,j|t)}{dt} = -\sum_{\mu=1}^{J} \delta_i^{\mu j}(t)N(i,j|t) + \sum_{\mu=1}^{J} \delta_i^{j\mu}(t)N(i,\mu|t) - \beta_i^j(t)N(i,j|t) - \alpha_i^j(t)N(i,j|t)$$

$$+ \begin{cases} 0 & , i = 0 \\ 2\sum_{\mu=1}^{J} \alpha_{i-1}^{\mu}(t)w_{i-1}^{j\mu}(t)N(i-1,\mu|t) & , i \geq 1 \end{cases},$$

$$(4.28)$$

with initial conditions $\forall j : N(0,j|0) = N_{\text{ini},j}, \forall i \geq 1 : N(i,j|0) = 0$, as in (4.14). A model out of this model class will be further discussed in the next chapter in Section 5.2.2.

Label-structured (L) Label-structured models have been introduced, before more general DL models were available, to allow for an explicit modeling of the label concentration and thus to provide a link to data from labeling experiments (Banks et al., 2010; Luzyanina et al., 2007b, 2009). An L model is given by a PDE

$$\frac{\partial n_L(x|t)}{\partial t} + \frac{\partial(\nu(x,t)\,n_L(x|t))}{\partial x} = -\beta(x,t)n_L(x|t) - \alpha(x,t)n_L(x|t)$$

$$+ 2\alpha(x,t)\int_{\mathbb{R}_+} p_{\text{lab}}(x-\chi|t)\,n_{L,\text{inhe}}(\chi|t)\,d\chi. \tag{4.29}$$

Instead of division- or/and cell type-dependent proliferation parameters, the division and death rates were originally chosen to depend (besides time-dependency) on the label concentration x. For the case of CFSE labeling, for which the L model was introduced, the second line of (4.29) further simplifies to $2\gamma\alpha(x,t)n_L(\gamma x|t)$, with a parameter $\gamma \in (1,2]$ to account for label dilution (Luzyanina et al., 2007b).

If the proliferation parameters are independent of the label concentration, $\forall x : \alpha(x,t) = \alpha(t), \beta(x,t) = \beta(t)$, then the L model can be derived from the DL model (or a CL model, or a DCL model) by marginalization over the division numbers i (or/and the cell types j, respectively). The overall label distribution of the DCL model, $m(x|t)$ as defined in (4.9), then equals the solution of the L model, $n_L(x|t)$.

Division-structured (D) Division-structured population models have been presented in numerous publication, for example, by Deenick et al. (2003); Ganusov et al. (2005); Gett and Hodgkin (2000); Lee and Perelson (2008); León et al. (2004); Luzyanina et al. (2007a); Revy et al. (2001). Generally, a D model is given by a system of ODEs, $\forall i \in \mathbb{N}_0$:

$$\frac{d\underline{N}(i|t)}{dt} = -\underline{\beta}_i(t)\underline{N}(i|t) - \underline{\alpha}_i(t)\underline{N}(i|t) + \begin{cases} 0 & , i = 0 \\ 2\underline{\alpha}_{i-1}(t)\underline{N}(i-1|t) & , i \geq 1 \end{cases}, \tag{4.30}$$

with initial conditions $\forall j : \underline{N}(0|0) = \underline{N}_{\text{ini}}, \forall i \geq 1 : \underline{N}(i|0) = 0$.

In case of one cell type, $J = 1$, the DC model trivially equals the D model. In case of more than one cell type, $J > 1$, the D model can be derived from the DC model my marginalization over the cell types. Then, the state variables $\underline{N}(i|t)$ of the D model are related to the state variables of the DC model as in (4.4) for $K = 1$, and the parameters of the D model are derived from the parameters of the DC model as

$$\underline{\alpha}_i(t) := \frac{\sum_{j=1}^{J} \alpha_i^j(t) N(i,j|t)}{\sum_{j=1}^{J} N(i,j|t)}, \quad \underline{\beta}_i(t) := \frac{\sum_{j=1}^{J} \beta_i^j(t) N(i,j|t)}{\sum_{j=1}^{J} N(i,j|t)}, \tag{4.31}$$

whereas it should be noted that these marginalized division-structured parameters depend on the values of the division- and cell type-structured state variables $N(i,j|t)$.

Cell type-structured (C) Cell type-structured models have been used, for example, to model multi-cell type populations comprising naive, experienced and activated T cells (De Boer and Perelson, 1995), or activated and memory T cells (De Boer et al., 2003a). The general form of a C model is given by a system of ODEs, $\forall j \in \{1, \ldots, J\}$:

$$\frac{d\overline{N}(j|t)}{dt} = - \sum_{\mu=1}^{J} \overline{\delta}^{\mu j}(t) \overline{N}(j|t) + \sum_{\mu=1}^{J} \overline{\delta}^{j\mu}(t) \overline{N}(\mu|t)$$

$$- \overline{\beta}^j(t) \overline{N}(j|t) - \overline{\alpha}^j(t) \overline{N}(j|t) + 2 \sum_{\mu=1}^{J} \overline{\alpha}^{\mu}(t) \overline{w}^{j\mu}(t) \overline{N}(\mu|t), \tag{4.32}$$

with initial conditions $\forall j : \overline{N}(j|0) = \overline{N}_{\text{ini},j}$.

The C model can be derived from the DC model my marginalization over the division numbers. Then, the state variables $\overline{N}(j|t)$ of the C model are related to the state variables of the DC model as given in (4.5) with $K = 1$, and the parameters of the C model are derived from the parameters of the DC model as

$$\overline{\delta}^{j_2 j_1}(t) := \frac{\sum_{i=0}^{\infty} \delta_i^{j_2 j_1}(t) N(i,j_1|t)}{\sum_{i=0}^{\infty} N(i,j_1|t)}, \quad \overline{\beta}^j(t) := \frac{\sum_{i=1}^{\infty} \beta_i^j(t) N(i,j|t)}{\sum_{i=1}^{\infty} N(i,j|t)},$$

$$\overline{\alpha}^j(t) := \frac{\sum_{i=1}^{\infty} \alpha_i^j(t) N(i,j|t)}{\sum_{i=1}^{\infty} N(i,j|t)}, \quad \overline{w}^{j_2 j_1}(t) := \frac{\sum_{i=1}^{\infty} \alpha_i^{j_1}(t) w_i^{j_2 j_1}(t) N(i,j_1|t)}{\sum_{i=0}^{\infty} \alpha_i^{j_1}(t) N(i,j_1|t)}. \tag{4.33}$$

Again, it has to be noted that these marginalized cell type-structured parameters depend on the division- and cell type-structured state variables $N(i,j_1|t)$. In this chapter, a C model will be used in Subsection 4.3.2 to link cell type dynamics to gene switch dynamics.

DsCL models with two time intervals (D_2CL)   Specifying a DsCL model for BrdU labeling usually requires two time intervals, $K = 2$, as will be elaborated in this work in Chapter 6. Consequently, the number of divisions is tracked by a two-dimensional division sequence vector, $\mathbf{i} = (i_1, i_2)$. To the best of our knowledge, this is the first model for BrdU-labeled proliferating populations that jointly considers division numbers in two time intervals, cell types, and the labeling properties of BrdU at the same time. This model, which we call a D_2CL model to denote the sequence of two time intervals, is given by the system of PDEs (4.1). To account for the specific labeling properties arising in BrdU experiments, a detailed derivation for modeling the BrdU-specific label distribution, $\int_{\mathbb{R}_+} p_{\mathrm{lab}}(x - \chi|t) \int_{\mathbb{R}_+} p_{\mathrm{inhe}}(\chi|\zeta, \mathbf{i} - \mathbf{e}_k^T) n_L(\zeta|t) \, d\zeta \, d\chi$, is presented within the model development in Chapter 6. Importantly, the label distribution derived in Chapter 6 as well as a more phenomenological variant presented previously (Schittler et al., 2013a) lead to models that can again be decomposed, as given by (4.12).

Division-, two intervals- and label-structured (D_2L)   A model structured according to division in two time intervals and label concentration has been introduced in Schittler et al. (2013a) to account, to the best of the author's knowledge as the first one, for divisions in two time intervals (uplabeling and delabeling) in BrdU experiments, and the full label distribution of BrdU concomitantly. In principal the model employed therein resembles the system of PDEs (4.27) but with the division sequence vector $\mathbf{i} = (i_1, i_2)$ replacing the division number i, and no degradation $\nu(x, t) = 0$ since BrdU is not degraded. In the model presented in Schittler et al. (2013a), the label distribution was approximated heuristically. In order to proceed from such a phenomenological description towards a more mechanistic description, we will develop a comprehensive model for BrdU labeling in Chapter 6.

Division-, two intervals- and cell type-structured (D_2C)   For the purpose of modeling BrdU-labeled proliferating cell populations, numerous models have been presented that are structured according to divisions in two time intervals and according to cell types, for example to investigate T cell populations (Ganusov and De Boer, 2013; Parretta et al., 2008), or hematopoietic stem cell populations (Wilson et al., 2008; Glauche et al., 2009; Bonhoeffer et al., 2000). Interestingly, the importance of considering multiple cell types was noticed earlier in models of BrdU-labeled populations (Bonhoeffer et al., 2000; Glauche et al., 2009) than for models of CFSE-labeled populations. Typical D_2C models distinguish, for example, resting and activated cells (Bonhoeffer et al., 2000; Wilson et al., 2008), naive and memory cells (Parretta et al., 2008), or more generally, fast and slow dividing cells (Ganusov and De Boer, 2013; Glauche et al., 2009).

Similarly to the corresponding models for one time interval, the D_2C model can be derived from the D_2CL model by marginalizing over the label concentration x, as in (4.3), or again if the decomposition (4.12) can be conducted.

Division- two intervals-structured (D_2)   The fact that BrdU labeling experiments usually consist of two time intervals with different labeling properties motivated to consider divisions in two subsequent time intervals already in early studies (Kiel et al., 2007). However, as mentioned above, when fitting models to data from BrdU labeling experiments, the D_2C models that additionally consider cell types have proven to be of advantage.

To sum up, it is seen that the DsCL model class presented here is quite general, in the sense that it covers a broad range of labeling experiment settings commonly used to study proliferating and multi-cell type populations. As discussed above, many available models are contained in this model class as special cases. Therefore, investigating the general DsCL model can yield powerful results for solving and analyzing a multitude of models describing cell populations.

In common settings of proliferation experiments, the specific properties of labeling techniques, such as CFSE or BrdU, can be exploited to decompose the specific model and thus reduce its complexity considerably, enabling efficient analyses and simulations. At the same time, the general DsCL model class provides, out of all here discussed models, the most highly resolved information, namely about cell types, division numbers in subsequent time intervals, and label concentration. Depending on the specific problem at hand, the state variables provided by the DsCL model class can be marginalized and matched to data.

4.3.2 Relationship to gene switch models

The cell type-structured models discussed above provide a description of multi-cell type dynamics on the population level. As this, they can be regarded as phenomenological descriptions of cell type transitions. A somewhat more mechanistic description of cell type transitions is offered by models of gene regulation that determines the cell type, as they were the focus of Chapters 2 and 3.

We will now investigate to what extent the cell type dynamics arising from a gene switch can be represented by a cell type-structured model (4.32). Connecting the gene switch models to the C models and utilizing the above discussed classification of the C models within the DsCL model class will illuminate the relationship between models of gene regulation on the single-cell level and the DsCL model class as a comprehensive class of population models.

The aim of this subsection is

- to show how cell type transitions induced by a gene switch may be translated into fluxes (parameters) in a corresponding C model, and

- to identify necessary conditions on the properties of gene switch dynamics and cell content segregation for generating cell type transitions.

For this purpose, we will derive a population balance equation (PBE) model from a gene switch model with deterministic and stochastic dynamics. Together with the regions of cell type-characteristic gene expression, we will use this PBE model to relate the emerging fluxes to the dynamics in a C model.

Derivation of a population balance equation with gene expression dynamics We now consider a cell population, in which each cell has intracellular gene expression including deterministic and stochastic dynamics (2.2). Furthermore, the cell population is assumed to be subject to the common processes of cell division and death. The resulting cell population dynamics can be written as a PBE, as pointed out in the following. The derivation of this PBE is on the basis of Ramkrishna (2000), however, in their work it was assumed that upon cell division, the newborn cells have exactly the same gene expression state as their mother cell.

To disentangle from this restrictive assumption, we incorporate a generalized partitioning of gene expression products in the style of Mantzaris (2006, 2007).

The state variable of a population balance model in general is the number density $n(z|t)$ of cells with a certain property z at time t (Fredrickson et al., 1967; von Foerster, 1959). As we are interested in the level of gene expression, we will define the variable $z \in \mathbb{R}_+^n$ as the expression levels of n genes or TRs. Let us first focus on the population-wide gene expression dynamics, and for the time being write unspecific birth and loss terms which will be replaced later by specific expressions. Let the gene expression dynamics be given by an SDE (2.2). The overall gene expression level of the population is then given by

$$\int_{\mathbb{R}_+^n} z\, n(z|t) dz. \tag{4.34}$$

This gene expression level changes due to

- the birth and death of cells with this value, $\text{birth}(n(z|t)) - \text{loss}(n(z|t))$, with the birth and loss term to be specified later, and

- the dynamics in z itself, which we express due to their partly stochastic nature via the expectation value, $\mathbb{E}\left(\frac{dz}{dt}\right) n(z|t)$.

With this, the gene expression in the overall population changes as

$$\int_{\mathbb{R}_+^n} z \frac{\partial}{\partial t} n(z|t) dz = \int_{\mathbb{R}_+^n} z \left(\text{birth}(n(z|t)) - \text{loss}(n(z|t))\right) dz + \int_{\mathbb{R}_+^n} \mathbb{E}\left(\frac{dz}{dt}\right) n(z|t) dz. \tag{4.35}$$

As shown in Appendix F in the style of Ramkrishna (2000), this yields for the gene expression dynamics in the PBE,

$$\frac{\partial}{\partial t} n(z|t) = -\nabla \left(f(z,u) n(z|t)\right) + \frac{\sigma^2}{2} \nabla^2 \left(g(z,u) n(z|t)\right) + \left(\text{birth}(n(z|t)) - \text{loss}(n(z|t))\right), \tag{4.36}$$

with divergence $\nabla := \partial/\partial z_1 + \ldots + \partial/\partial z_n$, and laplacian $\nabla^2 := \partial^2/\partial z_1^2 + \ldots + \partial^2/\partial z_n^2$. The drift term f and the diffusion term g therein originate from the SDE model (2.2).

Now, let us also specify the birth and loss terms. Loss of cells $n(z|t)$ occurs due to death with a death rate $\beta(z,t)$, and due to division with a division rate $\alpha(z,t)$. Thus, the loss term reads

$$\text{loss}(n(z|t)) = -\left(\alpha(z,t) + \beta(z,t)\right) n(z|t). \tag{4.37}$$

The appearance of newborn cells is a bit more complex: If a mother cell with gene expression state ζ divides, the cell will give rise to two daughter cells, whereby the gene expression level may be partitioned to the daughter cells in a nontrivial way. Let this probability, that mother cells $n(\zeta|t)$ yield daughter cells $n(z|t)$, be represented by the general probability density $p_{\text{birth}}(z|\zeta)$ (Fredrickson et al., 1967; Mantzaris, 2006, 2007). Cells from $n(\zeta|t)$ divide with division rate $\alpha(\zeta,t)$, thereby yielding two new daughter cells per division event, such that we can write the birth term as

$$\text{birth}(n(z|t)) = 2 \int_{\mathbb{R}_+^n} \alpha(\zeta,t) p_{\text{birth}}(z|\zeta) n(\zeta|t) d\zeta. \tag{4.38}$$

By inserting the obtained birth and loss terms, (4.37) and (4.38), into the PBE (4.36), we obtain the PBE including cell division and death,

$$
\begin{aligned}
\frac{\partial}{\partial t} n(z|t) = & - \nabla \left(f(z,u)n(z|t) \right) + \frac{\sigma^2}{2} \nabla^2 \left(g(z,u)n(z|t) \right) \\
& - \left(\alpha(z,t) + \beta(z,t) \right) n(z|t) + 2 \int_{\mathbb{R}_+^n} \alpha(\zeta,t)p_{\text{birth}}(z|\zeta)n(\zeta|t)d\zeta,
\end{aligned}
\tag{4.39}
$$

with $n(z|t) \in \mathbb{R}_+$ the number density of cells with gene expression level z at time t, u the input, $f(z,u)$ the drift term, $g(z,u)$ the diffusion term, $\alpha(z,t) \in \mathbb{R}_+$ and $\beta(z,t) \in \mathbb{R}_+$ division and death rates, and $p_{\text{birth}}(z|\zeta)$ the probability density describing the segregation of the gene expression products to daughter cells.

The obtained PBE (4.39) needs to be accompanied by suitable initial conditions, $n(z|0) = n_{\text{ini}}(z)$, as well as boundary or regularity conditions. Commonly, such regularity conditions are imposed stating that there is no flux of cells out of the region of possible values (Fredrickson et al., 1967; Mantzaris, 2006; Stamatakis, 2010). In a biologically meaningful model, the gene expression level z cannot become negative nor infinite, thus there will be boundaries $\partial\Omega := \{z | \exists i \in \{1, \ldots, n\} : z_i = 0 \vee z_i \to \infty\}$ with

$$
\forall z \in \partial\Omega : \left(f_i(z,u) - \frac{\sigma^2}{2} \frac{\partial}{\partial z_i} g_i(z,u) \right) n(z|t) = 0.
\tag{4.40}
$$

Let us now employ the obtained PBE model along with a partitioning of the state space. The gene expression state $z \in \mathbb{R}_+^n$ determines the cell type $j \in \{1, \ldots, J\}$, based on the attractor regions of the stable steady states $z^{*,j}$, according to

$$
j \Leftrightarrow z \in \Omega_j := \{z_0 | \lim_{t \to \infty} z(t, z_0) = z^{*,j}\},
\tag{4.41}
$$

wherein $z(t, z_0)$ is the solution of the deterministic version of the gene switch dynamics given by an ODE model (2.1) in the variable z, with $z(0) = z_0$.

For simplicity of notation, let us in the following consider a one-dimensional gene switch, $n = 1$, with two stable steady states $z^{*,1}$ and $z^{*,2}$, separated by an unstable steady state z^c, with $z^{*,1} < z^c < z^{*,2}$. Although the state space could be partitioned in general by any arbitrary intervals, we will exploit the properties of the gene switch dynamics to simplify the fluxes between the cell types. We will also consider stochastic dynamics as well as inputs, but the steady states are defined from the deterministic dynamics under no input, $u = 0$. Due to (4.41), such a gene switch partitions the state space into two domains $\Omega_1 = [0, z^c)$ and $\Omega_2 = (z^c, \infty)$, thus it gives rise to $J = 2$ cell types. The results obtained in the following can be transferred directly to gene switches with n dimensions, and/or with more than two steady states and thus a state space partitioning according to J cell types. Also for simplicity of notation, we will assume that division and death rates are independent of the gene expression, $\alpha(z,t) = \alpha(t)$ and $\beta(z,t) = \beta(t)$. The results can be generalized to gene expression level-dependent division and death rates by simply replacing the rates by their integrals over the respective domain.

Based on the definition of cell types in terms of their gene expression (4.41), we will denote the numbers of cells of cell type $j = 1$ and cell type $j = 2$ by the variables

$$
N_j(t) := \int_{\Omega_j} n(z|t)dz.
\tag{4.42}
$$

Let us now write the dynamics of the number of cells, by integrating the dynamics given by the PBE model (4.39) over the respective domains as given by (4.42). Substituting the terms at $z = 0$ and $z \to \infty$ by the regularity condition (4.40), and the integrals by the number of cells (4.42), this yields

$$
\frac{dN_1(t)}{dt} = - f(z^c, u)n(z^c|t) + \frac{\sigma^2}{2} \frac{\partial(g(z, u)n(z|t))}{\partial z}\Big|_{z^c}
$$
$$
- (\alpha(t) + \beta(t))\, N_1(t) + 2\alpha(t) \int_{\mathbb{R}_+} \int_{\Omega_1} p_{\text{birth}}(z|\xi)dz\, n(\xi|t)d\xi
$$
(4.43)

$$
\frac{dN_2(t)}{dt} = f(z^c, u)n(z^c|t) - \frac{\sigma^2}{2} \frac{\partial(g(z, u)n(z|t))}{\partial z}\Big|_{z^c}
$$
$$
- (\alpha(t) + \beta(t))\, N_2(t) + 2\alpha(t) \int_{\mathbb{R}_+} \int_{\Omega_2} p_{\text{birth}}(z|\xi)dz\, n(\xi|t)d\xi
$$
(4.44)

Therein, in both equations the first line gives the gene switch dynamics, including deterministic dynamics, stochastic dynamics, and inputs affecting gene expression. The second line gives, in the first term, the loss of cells, and in the second term the appearance of new cells due to cell division, possibly with cell type transition. In the following, we may refer to this set of equations (4.43), (4.44) as a gene switch-structured population model.

Identification with a cell-type structured model The PBE model (4.39) eludes a straight-forward solution and simulation, and other authors that have used similar PBE models have pursued different approaches to rewrite or approximate the PBE model (Mantzaris, 2006, 2007; Shu et al., 2012; Stamatakis, 2010). However, our aim here is not to obtain the dynamics themselves, which is suitably covered for example by an ensemble model of SDEs as presented in Chapter 2. Instead, we aim to deduce about *properties of the population-level dynamics*, arising from the gene switch, with respect to fluxes between cell types. Based on a PBE model we now show how to translate this into a C model. Due to its nature as an ODE system, it can be analyzed and solved clearly more efficiently than the PBE model (4.39) itself.

The dynamics in a cell population with two cell types, $J = 2$, may as well be represented by a C model (4.32). The C model of such a cell population reads

$$
\frac{d\overline{N}(1|t)}{dt} = \overline{\delta}^{12}(t)\overline{N}(2|t) - \overline{\delta}^{21}(t)\overline{N}(1|t)
$$
$$
- (\overline{\alpha}^1(t) + \overline{\beta}^1(t))\overline{N}(1|t) + 2\overline{w}^{11}(t)\overline{\alpha}^1(t)\overline{N}(1|t) + 2\overline{w}^{12}(t)\overline{\alpha}^2(t)\overline{N}(2|t)
$$
(4.45)

$$
\frac{d\overline{N}(2|t)}{dt} = \overline{\delta}^{21}(t)\overline{N}(1|t) - \overline{\delta}^{12}(t)\overline{N}(2|t)
$$
$$
- (\overline{\alpha}^2(t) + \overline{\beta}^2(t))\overline{N}(2|t) + 2\overline{w}^{22}(t)\overline{\alpha}^2(t)\overline{N}(2|t) + 2\overline{w}^{21}(t)\overline{\alpha}^1(t)\overline{N}(1|t)
$$
, (4.46)

where in each equation, the first line reflects the fluxes due to cell type transitions (without division). In the second line, the first term denotes the loss of cells, the second term represents the appearance of cells due to cell division (without cell type transition), and the third term denotes the appearance of cells due to cell type transitions upon division.

Now, let us identify the numbers of cells in the C model (4.45), (4.46) with the numbers of cells in the gene switch structured population model (4.43), (4.44), for $j = 1, 2$:

$$\overline{N}(j|t) = N_j(t) = \int_{\Omega_j} n(z|t)dz, \tag{4.47}$$

with Ω_j being the attractor basins according to (4.41). From this, the fluxes in the gene switch structured population model (4.43), (4.44) can be matched to the fluxes in the C model (4.45), (4.46). We will conduct this matching and deduce relationships between the dynamics in the two compared models now subsequently for the individual fluxes of cell type transitions (without division), loss of cells, and birth of cells (possibly with cell type transitions upon division).

Cell type transitions without division   First, let us investigate the fluxes of cell type transitions (without cell division). Therefore we match them to the population's gene switch dynamics,

$$\overline{\delta}^{j_2 j_1}(t)\overline{N}(j_1|t) = \max\left\{0, \int_{\Omega_{j_1}} \left(-(f(z, u)n(z|t)) + \frac{\sigma^2}{2}\frac{\partial}{\partial z}(g(z, u)n(z|t))\right) dz\right\}, \tag{4.48}$$

wherein taking the maximum means that the positive net flux between the cell types is considered. As derived in Appendix G, the possibility of having nonzero fluxes $\overline{\delta}^{21}(t)\overline{N}(1|t)$ or $\overline{\delta}^{12}(t)\overline{N}(2|t)$ is constrained by the gene switch dynamics as follows:

- For deterministic dynamics without input, that is $u = 0$ and $g(z, u) = 0$, it must hold that

$$\begin{aligned} \overline{\delta}^{21}(t)\overline{N}(1|t) &= 0 \\ \overline{\delta}^{12}(t)\overline{N}(2|t) &= 0. \end{aligned} \tag{4.49}$$

  This means that no cell type transitions can be induced by purely deterministic gene switch dynamics if no input is applied on the system.

- For deterministic dynamics with input, $u \neq 0$ but $g(z, u) = 0$, it is

$$\begin{aligned} \overline{\delta}^{21}(t)\overline{N}(1|t) &= \max\left\{0, \Delta_f(z^c, u)n(z^c|t)\right\} \\ \overline{\delta}^{12}(t)\overline{N}(2|t) &= \max\left\{0, -\Delta_f(z^c, u)n(z^c|t)\right\}, \end{aligned} \tag{4.50}$$

  with $\Delta_f(z, u) := f(z, u) - f(z, 0)$ in order to separate the dynamics induced by a nonzero input. From this we see that, depending on the sign of $\Delta_f(z^c, u)$ and for $n(z^c|t) \neq 0$, in one of the two directions there can be a nonzero flux. However, the flux will be unidirectional, such that a deterministic gene switch with input can not induce cell type transitions in both directions. Moreover, if no further source of cells supplies the cell type from which the flux is coming, the flux will vanish over time, due to its unidirectionality, such that there cannot be sustained cell type transitions.

- For stochastic dynamics, $g(z, u) \neq 0$ (but set $u = 0$), it is

$$\begin{aligned} \overline{\delta}^{21}(t)\overline{N}(1|t) &= \max\left\{0, -\frac{\sigma^2}{2}\frac{\partial(g(z, u)n(z|t))}{\partial z}\bigg|_{z^c}\right\} \\ \overline{\delta}^{12}(t)\overline{N}(2|t) &= \max\left\{0, \frac{\sigma^2}{2}\frac{\partial(g(z, u)n(z|t))}{\partial z}\bigg|_{z^c}\right\}. \end{aligned} \tag{4.51}$$

From this it is seen that fluxes may occur in both directions, induced by the stochastic gene switch dynamics. Since the direction of cell type transition is not fixed, but may change according to changes in $n(z^c|t)$ (respectively, its gradient over z), there may be an ongoing exchange of cells between both cell types, and thus one can not rule out that there are sustained cell type transitions.

These investigations of cell type transition fluxes arising from gene switch dynamics bring us to the following conclusion: Deterministic dynamics without input cannot induce cell type transitions. Deterministic dynamics with input can induce cell type transitions, though only unidirectional. Stochastic dynamics can induce sustained cell type transitions in both directions.

Loss of cells To evaluate the loss of cells, the fluxes are matched by

$$(\overline{\alpha}^j(t) + \overline{\beta}^j(t))\overline{N}(j|t) = (\alpha(t) + \beta(t))\,N_j(t). \tag{4.52}$$

Due to (4.42), this states that $(\overline{\alpha}^j(t) + \overline{\beta}^j(t)) = (\alpha(t) + \beta(t))$. This also includes that the sums of division and death rates need to be equal for all cell types j, which follows from the assumption that division and death rates are independent of the gene expression level.

Birth of cells with and without cell type transition Regarding the birth of cells which do not change their cell type upon division, the fluxes are given by

$$2\overline{w}^{j_1 j_1}(t)\overline{\alpha}^{j_1}(t)\overline{N}(j_1|t) = 2\alpha(t) \int_{\Omega_{j_1}} \int_{\Omega_{j_1}} p_{\text{birth}}(z|\zeta)d\zeta\, n(\zeta|t)dz. \tag{4.53}$$

Similarly, for the cell divisions with simultaneous cell type transitions from j_1 to j_2, the fluxes are given by

$$2\overline{w}^{j_2 j_1}(t)\overline{\alpha}^{j_1}(t)\overline{N}(j_1|t) = 2\alpha(t) \int_{\Omega_{j_2}} \int_{\Omega_{j_1}} p_{\text{birth}}(z|\zeta)d\zeta\, n(\zeta|t)dz. \tag{4.54}$$

Given that cells of type j_1 do divide, this gives a necessary condition for nonzero transition-upon-division fluxes from j_1 to j_2: These fluxes are only possible if there exist some $\zeta \in \Omega_{j_1}, z \in \Omega_{j_2}$ for which $p_{\text{birth}}(z|\zeta) \neq 0$, that is, if the probability that a cell with gene expression level $\zeta \in \Omega_{j_1}$ yields a daughter cell with gene expression level $z \in \Omega_{j_2}$ is nonzero.

Let us briefly conclude from the investigations in this section. As only the cell type transition rates $\overline{\delta}^{j_2 j_1}(t)$ depend on the gene switch dynamics, the dynamical properties of the genetic switch predetermine the properties of cell type transitions. The presence of stochasticity in the gene switch dynamics is thereby a necessary condition for bidirectional transition fluxes between the cell types. In contrast, the cell type transition-upon-division probabilities $w^{j_2 j_1}(t)$ depend on the partitioning of gene expression products from mother to daughter cells, $p_{\text{birth}}(z|\zeta)$. The existence of nonzero probabilities $p_{\text{birth}}(z|\zeta)$ for z and ζ from distinct cell type domains has been identified as a necessary condition for type transitions-upon-division between these cell types.

4.4 Summary and discussion

In this chapter, we presented a novel and general model class for labeled proliferating multi-cell type populations and thereby solved Problem 4.1. The model class is general in the sense that it accounts for division numbers, cell types, label concentration, and time intervals with changing labeling conditions. Such models are needed, because division- and/or cell type-dependent proliferation properties as found in the majority of cell populations can only be uncovered by the help of mathematical models. Furthermore, the label concentration is the variable detected as the measurement output, and time intervals with changing labeling conditions affect the label dynamics. The model class is general also because, importantly, it does not make specific assumptions on the labeling technique, but covers diverse common experimental settings. Consequently, a multitude of existing population models falls into the here presented class, and they become amenable to the developed simulation and analysis tools.

Representing the cell type by a discrete variable results in a model of clearly lower complexity than, for example, using continuous variables such as the cell type-determining gene expression. The latter approach results in PBE models with dynamics also in the gene expression variables, for which the solution usually becomes computationally demanding or impossible at all (Mantzaris, 2006, 2007; Sidoli et al., 2006; Shu et al., 2012; Stamatakis, 2010). If the detailed gene expression dynamics are not known and rather the cell type transition fluxes are of interest, the here presented cell type-structured population models with their drastically lower complexity may offer a suitable approach.

A major contribution of this generalized modeling approach is clearly that methodological questions may be posed, and modeling methods may be developed, on this general level, with results addressing a broad range of models independently of the application context. For a particular application, a suitable model can then be derived by specifying the model equations according to the experimental settings. As will be seen in such specified models for CFSE and BrdU labeling in Chapters 5 and 6, for common settings the model equations, given by a set of coupled PDEs, can be decomposed into a set of ODEs and a set of distributions. This tremendously reduces the complexity and thus enables much more efficient solution methods. It is therefore worth, as we did in this chapter, to further study this subset of decomposable models, in order to develop advantageous analysis and simulation tools. Not only can analytical solutions allow for the understanding of dynamical behavior without the need for exhaustive simulation-based studies, but also numerical solutions can be made arbitrarily close to the exact solution via the truncation error bound derived in this chapter.

As this general model class comprises many established models as special cases, we dedicated a comprehensive discussion to the classification and relationship of these models within the inroduced model class. Moreover, we showed how the cell type structure of this model class offers a link to gene switch models. The gene switch dynamics relate to parameters in a cell type-structured population model, whereas in turn necessary conditions for certain population dynamics were derived on the gene switch dynamics. Cell type-structured population models are a subclass of the general model class presented in this chapter, and they were put into relation to other models within this class. We have opened the link between gene switch dynamics and population models by investigating gene switch-induced transition fluxes in a cell type-structured model. Similar analyses may also be performed for population models structured not only by cell type, but also by division numbers/sequences, and/or label concentration, whereas the notation would become more complex.

Chapter 5

Modeling of CFSE-labeled multi-cell type populations

This chapter is dedicated to the derivation of a specific model for multi-cell type populations labeled with Carboxyfluorescein succinimidyl ester (CFSE). Emanating from the general model class for proliferating multi-cell type populations in labeling experiments, which was elaborated in the previous chapter, we exploit the properties of the CFSE labeling technique in order to specify a suitable model. Section 5.1 provides a brief background on CFSE labeling and available models, whereafter we formulate the problem addressed in this chapter. In Section 5.2, we develop the model as well as efficient solution and analysis approaches. Importantly, the labeling properties of CFSE lead to a model which allows for a decomposition approach, which simplifies the solution of the model PDE system drastically. The obtained model and accompanying approaches are applied in Section 5.3 to study the example of a CFSE-labeled proliferating stem cell population comprising two cell types. Finally, we conclude with a short summary and discussion in Section 5.4.

This chapter is partly based on Schittler et al. (2011, 2012).

5.1 Background and problem formulation

5.1.1 CFSE labeling experiments

To study cell proliferation, a commonly applied experimental technique is labeling with Carboxyfluorescein succinimidyl ester (CFSE). CFSE is an intracellular fluorescent dye that is brought into the cells in a chemically slightly different form, namely Carboxyfluorescein diacetate succinimidyl ester (CFDA-SE), which is highly cell permeable. Once inside the cell, CFDA-SE is converted in the cytoplasm into CFSE, which in turn covalently binds to intracellular proteins and is not noticably cell permeable, meaning that the dye remains in the cell. Upon cell division, the concentration of CFSE is approximately halved in the daughter cells (Lyons and Parish, 1994; Lyons, 2000). The concentration of CFSE in each individual cell can be measured by flow cytometry, which allows for conclusions about the number of undergone cell divisions at the respective time.

Detailed reports on the usage of CFSE for cell proliferation studies can be found, for example, by Hawkins et al. (2007); Lyons and Parish (1994); Lyons (2000). Importantly, cell proliferation has been shown to be unaffected by CFSE at least when administered at moderate levels (Lyons and Parish, 1994; Matera et al., 2004). A general discussion of the derived data type, termed as binned snapshot data, is given in Hasenauer (2012). The decay of the label CFSE has been suggested to be determined by the degradation of proteins in the cytoplasm to which CFSE is bound, and the derivation of suitable quantitative descriptions for the CFSE decay process has been the subject of several studies (Banks et al., 2011, 2013; Hasenauer, 2012).

Since its establishment, CFSE has been used for proliferation studies on cell systems as diverse as stem cells (Groszer et al., 2001; Prudhomme et al., 2004; Urbani et al., 2006), lymphocytes (Lyons, 2000), leukemic cells (Holyoake et al., 1999), and bacteria (Ueckert et al., 1997), to name just a few examples.

5.1.2 Available models for CFSE-labeled cell populations

In the first approaches to quantitatively interpret CFSE data, the number of cells within each individual peak, each corresponding to a different division number, was counted (Hawkins et al., 2007; Lyons and Parish, 1994; Lyons, 2000; Nordon et al., 1999; Wells et al., 1997). This approach is possible with CFSE data because usually cells of the same division number exhibit a similar label concentration within a narrow range, and thus separated peaks can be detected which correspond to separate division numbers. Although this simple method may seem tempting, it suffices from several drawbacks: Even if separate peaks are detectable, they overlap to a considerable extent, and the peaks get closer with increasing division numbers until they approach the autofluorescence. The required steps of deconvolving cell numbers thereby introduce additional error sources. Finally, more complex dynamics such as division- or time-dependent proliferation rates are not considered.

In order to establish more dynamical modeling approaches, numerous division-structured population models have been proposed and used for CFSE labeling (Deenick et al., 2003; Ganusov et al., 2005; Gett and Hodgkin, 2000; Lee and Perelson, 2008; León et al., 2004; Luzyanina et al., 2007a; Revy et al., 2001). These models allow for division-dependent parameters as well as the quantitative description of the number of cells with certain division number, and have also been discussed in Chapter 4. Similarly, modeling cell proliferation as branching processes also imposes a division-structure on the modeled population (Yates et al., 2007; Miao et al., 2012). However, division-structured population models do not account for explicit label intensities, thus still requiring error-prone derivation of cell numbers from measurements of label intensity.

As a first step towards modeling the label intensity, the mean fluorescence intensity has been introduced (Asquith et al., 2006), and been used to estimate proliferation rates of lymphocytes in sheep (Debacq et al., 2006; Florins et al., 2006). The link to the full label intensity was established by label-structured population models (Banks et al., 2010, 2011; Luzyanina et al., 2007b, 2009), which have also been illuminated in Chapter 4. Finally, the generalizing division- and label-structured population models introduced recently (Hasenauer et al., 2012a,b; Schittler et al., 2011; Thompson, 2011) which combine advantages of the division-structured and the label-structured population models, have emerged from the model development for CFSE labeling.

5.1.3 Problem formulation

Numerous mathematical models for CFSE-labeled proliferating populations are available, but none of them accounts for multiple cell types besides the crucial properties of division numbers and label concentration. As outlined in the previous chapter in Section 4.1, however, the consideration of cell types with distinct proliferation properties is of high relevance. Therefore, the current chapter is dedicated to the following problem:

Problem 5.1 (*Modeling of CFSE-labeled proliferating multi-cell type populations*) *Given the general model for labeled proliferating multi-cell type populations, and the labeling properties of CFSE, develop an efficient multi-cell type population model for CFSE labeling.*

In the following, we will make use of the general DsCL model class developed in Chapter 4 and from that specify a model for a CFSE-labeled proliferating multi-cell type population. As will be seen, by exploiting the specific properties of CFSE labeling, we can derive an approach to efficiently analyze and simulate the model.

5.2 Specific model for CFSE-labeled multi-cell type populations

Let us first specify the model, and then look at its solution and analysis.

5.2.1 Model specification

In order to derive a model specifically tailored to CFSE labeling, out of the DsCL class, we exploit the properties of the CFSE labeling technique and translate them into mathematical expressions.

First we will consider the decay process of CFSE to specify the decay term, $\nu(x,t)$. As mentioned in Section 5.1.1, CFSE is bound to intracellular proteins which are due to their protein species-specific degradation rates. The degradation rates of each protein species can usually be assumed to be linear with a degradation constant. It has been suggested both from the biochemical processes of the degradation of protein-CFSE complexes, as well as from comparison of different degradation models, that the decay term for CFSE is best described by a sum of degradation processes (Banks et al., 2013) or, more precisely, by a sum of linear degradation processes (Hasenauer, 2012). These in turn can be written as a degradation process linear in x, and with time-dependent degradation constant $k(t)$, as argued in detail in Hasenauer (2012),

$$\nu(x,t) = -k(t)x, \tag{5.1}$$

where the degradation rate $k(t)$ can be expressed in dependence of the individual protein degradation constants (Hasenauer, 2012). This specification of the decay term, (5.1), will be an important property for the decomposition of the model equations later in this chapter.

Next, let us specify the label uptake and label inheritance from mother to daughter cells. In CFSE labeling experiments, the label is administered and incorporated at the beginning of the experiment, $t = 0$. After this, the labeling conditions remain unchanged. This means that a

CFSE labeling experiment can be described with $K = 1$ by one time interval, $(0, T_1]$. Furthermore, the label incorporation at the beginning can be reflected by an initial label distribution in the yet undivided cells,

$$i = 0 : \ n(x, 0, j|0) = n_{\text{ini},j}(x)$$
$$\forall i \neq 0 : \ n(x, i, j|0) = 0. \tag{5.2}$$

In the following, it is assumed that the label incorporation is independent of the cell type, which is a plausible assumption,

$$n_{\text{ini},j}(x) = N_{\text{ini},j} \cdot p_{\text{ini}}(x). \tag{5.3}$$

After the initial administration of CFSE, it is washed out from the medium, such that no label is incorporated in the remainder of the experiment, $\forall t > 0$:

$$p_{\text{lab}}(x|t) = \delta(x). \tag{5.4}$$

Upon cell division, the content of CFSE label is distributed approximately equally to both daughter cells (Lyons and Parish, 1994; Lyons, 2000). This means that a mother cell with label content x' will produce two new daughter cells with label content $x = \frac{1}{\gamma}x'$, with γ a factor for the label dilution due to cell division. Since actually the label content can be assumed to be halved when inherited to a daughter cell, the label dilution factor can usually be set $\gamma = 2$, but was introduced with the more flexible value $\gamma \in (1, 2]$ in previous models (Banks et al., 2010; Luzyanina et al., 2007a; Schittler et al., 2011). With this, the label inheritance can be expressed, with using the sifting property of the delta distribution, by $\forall i, j, t$:

$$p_{\text{inhe}}(x|i - 1, j, t) = \int_{\mathbb{R}_+} \delta(x - \frac{1}{\gamma}\chi)p(\chi|i - 1, t)d\chi$$
$$= \int_{\mathbb{R}_+} \gamma\delta(\gamma x - \chi)p(\chi|i - 1, t)d\chi \tag{5.5}$$
$$= \gamma p(\gamma x|i - 1, t).$$

The two integrals in the general DsCL model (4.1) that reflect the label inheritance and label uptake then simplify, inserting (5.4) for $p_{\text{lab}}(x|t)$ and (5.5) for $p_{\text{inhe}}(x|i - 1, j, t)$, and using the sifting property of the delta distribution, to

$$\int_{\mathbb{R}_+} p_{\text{lab}}(x - \chi|t)p_{\text{inhe}}(\chi|i - 1, \mu, t)d\chi$$
$$= \int_{\mathbb{R}_+} \delta(x - \chi)\gamma p(\gamma\chi|i - 1, t)d\chi = \gamma p(\gamma x|i - 1, t) \tag{5.6}$$

Exploiting these properties of CFSE labeling experiments, we can now write a model specifically tailored to CFSE-labeled multi-cell type populations, using the previously introduced DsCL model class and the specifications (5.1)–(5.6). The resulting model, representing the dynamics of $n(x, i, j|t)$, is given by a system of PDEs, $\forall i \in \mathbb{N}_0, \forall j \in \{1, \ldots, J\}$:

$$
\begin{aligned}
\frac{\partial n(x, i, j|t)}{\partial t} - k(t) \frac{\partial(x\, n(x, i, j|t))}{\partial x} = & \\
& - \sum_{\mu=1}^{J} \delta_i^{\mu j}(t) n(x, i, j|t) + \sum_{\mu=1}^{J} \delta_i^{j\mu}(t) n(x, i, \mu|t) \\
& - \beta_i^j(t) n(x, i, j|t) - \alpha_i^j(t) n(x, i, j|t) \\
& + \begin{cases} 0 & , i = 0 \\ 2\gamma \sum_{\mu=1}^{J} \alpha_{i-1}^{\mu}(t) w_{i-1}^{j\mu}(t) n(\gamma x, i-1, \mu|t) & , i \geq 1 \end{cases}
\end{aligned} \tag{5.7}
$$

with initial conditions

$$
n(x, 0, j|0) = n_{\text{ini},j}(x), \; \forall i \geq 1 : n(x, i, j|0) = 0. \tag{5.8}
$$

In the following, we will first prove that this model can be decomposed and thus be solved efficiently, and then use this model to study an exemplary CFSE-labeled stem cell population.

5.2.2 Model solution and analysis

The model for CFSE-labeled proliferating multi-cell type populations, as given by (5.7), is a system of coupled PDEs. The solution of such a PDE system requires sophisticated solution strategies involving numerical approximations and a high computational demand. Thanks to the properties of CFSE experiments as introduced in Subsection 5.2.1, the decomposition approach defined in the previous chapter by (4.12) can be exploited. Thereby the model becomes amenable to the solution methods that were presented in Section 4.2.2. In this section, we will first prove the decomposition, and then show how the overall solution can be obtained from the decomposed solutions.

Decomposability of the model The fact that the model (5.7) can be decomposed as in (4.12) as well as the emerging solution are captured by the following theorem:

Theorem 5.1 *The model (5.7) fulfills (4.11) and its solution is given by*

$$
\begin{aligned}
n(x, i, j|t) &= N(i, j|t) p(x|i, t) \\
&= N(i, j|t) \gamma^i e^{\int_0^t k(\tau) d\tau} p_{\text{ini}}(\gamma^i e^{\int_0^t k(\tau) d\tau} x),
\end{aligned} \tag{5.9}
$$

wherein $N(i, j|t)$ is the solution of the system of ODEs $\forall i \in \mathbb{N}_0, j \in \{1, \ldots, J\}$:

$$
\begin{aligned}
\frac{dN(i, j|t)}{dt} = & - \sum_{\mu=1}^{J} \delta_i^{\mu j}(t) N(i, j|t) + \sum_{\mu=1}^{J} \delta_i^{j\mu}(t) N(i, \mu|t) \\
& - \beta_i^j(t) N(i, j|t) - \alpha_i^j(t) N(i, j|t) \\
& + \begin{cases} 0 & , i = 0 \\ 2 \sum_{\mu=1}^{J} \alpha_{i-1}^{\mu}(t) w_{i-1}^{j\mu}(t) N(i-1, \mu|t) & , i \geq 1 \end{cases}
\end{aligned} \tag{5.10}
$$

with initial conditions

$$\forall j : N(0, j|0) = N_{\text{ini},j}, \forall i \geq 1 : N(i, j|0) = 0. \tag{5.11}$$

The quantities $N(i, j|t)$ are the numbers of cells in each subpopulation (4.3), and the quantities $p(x|i, t)$ are the label densities in each subpopulation (4.8).

Proof First, we prove that the model (5.7) fulfills (4.11). By (5.6) we already have rewritten the integral in (4.11), such that it only remains to show that the obtained expression in (5.6) indeed equals $p(x|i, t)$. Let us therefore solve the set of PDEs (4.15) which, with the decay term (5.1) for CFSE, becomes $\forall i \in \mathbb{N}_0$:

$$\frac{\partial p(x|i, t)}{\partial t} - k(t) \frac{\partial (x \, p(x|i, t))}{\partial x} = 0. \tag{5.12}$$

Its initial conditions can be obtained by inserting the combined label uptake and inheritance (5.6) during the experiment, $t \in (0, T_1]$, into (4.16),

$$p(x|i, 0) \overset{(4.16)}{=} \int_{\mathbb{R}_+} p_{\text{lab}}(x - \chi|T_1) p_{\text{inhe}}(\chi|i - 1, \mu, 0) d\chi \overset{(5.6)}{=} \gamma p(\gamma x|i - 1, 0). \tag{5.13}$$

Then, applying this recursively until one arrives at $i = 0$, (5.13) becomes

$$p(x|i, 0) = \gamma p(\gamma x|i - 1, 0) = \gamma^i p(\gamma^i x|0, 0) = \gamma^i p_{\text{ini}}(\gamma^i x). \tag{5.14}$$

The set of PDEs (5.12) with initial conditions (5.14) is a set of linear PDEs and has been shown by Hasenauer et al. (2012a) to have the solution

$$p(x|i, t) = \gamma^i e^{\int_0^t k(\tau)d\tau} p_{\text{ini}}(\gamma^i e^{\int_0^t k(\tau)d\tau} x). \tag{5.15}$$

From this it can be seen that indeed $\forall t$ it holds that $\forall i \in \mathbb{N}_0$:

$$p(x|i, t) = \gamma p(\gamma x|i - 1, t). \tag{5.16}$$

With this, it is proven that (4.11) is fulfilled. Using the decomposition (4.12) and the analytical solution of the PDE (5.15), one obtains the solution of the model as given in (5.9).

With this, Theorem 5.1 is proven. □

The decomposition offers a powerful tool for analyzing and simulating the model: The number of cells, $N(i, j|t)$, is obtained by solving the system of ODEs (5.10). This system can be solved either numerically or even analytically, as outlined in the previous Chapter. The probability densities $p(x|i, t)$ are simply obtained by calculating (5.15). From these two parts, the overall solution $n(x, i, j|t) = N(i, j|t)p(x|i, t)$ can be reassembled.

Overall model solution Following the decomposition approach which yielded the solution (5.9), it becomes visible that the problem of solving the model (5.7) reduces to solving the system of ODEs (5.10). How to solve this system of ODEs in order to derive $N(i, j|t)$ has

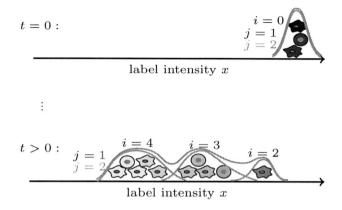

Figure 5.1: Composition of the population's overall label distribution, $m(x|t)$, from the label intensities of individual cells, of distinct cell types j and distinct division numbers i. Exemplarily, a population is illustrated with two distinct cell types, represented by star-shaped ($j = 1$) and round ($j = 2$) cells, respectively. At the beginning of the labeling experiment, $t = 0$, there are only undivided cells (denoted by division number $i = 0$), and irrespective of the cell type, all cells have approximately the same label intensity, up to uncertainties from the labeling process. After some time has passed, $t > 0$, the cells have undergone distinct numbers of divisions, for example, some cells have divided twice ($i = 2$), others three or four times ($i = 3, 4$). Upon division, the label content is halved to each daughter cell such that cells of higher division numbers will have a lower label intensity x. The label intensities of individual division numbers, $\underline{n}(x, i|t)$, are denoted by grey lines. The overall label intensity of the population, $m(x|t)$, is illustrated by the green line and gives the measurement output.

been treated in detail in the previous chapter, see Section 4.2.2. The overall label distribution, which is the variable to be matched to measurement output, is given by

$$
\begin{aligned}
m(x|t) &= \sum_{i=0}^{\infty} \sum_{j=1}^{J} n(x, i, j|t) \\
&= \sum_{i=0}^{\infty} \gamma^i e^{\int_0^t k(\tau)d\tau} p_{\text{ini}}(\gamma^i e^{\int_0^t k(\tau)d\tau} x) \sum_{j=1}^{J} N(i, j|t).
\end{aligned}
\tag{5.17}
$$

However, since the infinite sum cannot be evaluated in general, one may exploit the truncated ODE system as introduced before in Section 4.2.2. The overall label distribution is then approximated as

$$
\begin{aligned}
\hat{m}_S(x|t) &= \sum_{i=0}^{S-1} \sum_{j=1}^{J} n(x, i, j|t) \\
&= \sum_{i=0}^{S-1} \gamma^i e^{\int_0^t k(\tau)d\tau} p_{\text{ini}}(\gamma^i e^{\int_0^t k(\tau)d\tau} x) \sum_{j=1}^{J} N(i, j|t),
\end{aligned}
\tag{5.18}
$$

with S the number of divisions considered in the truncated ODE system. We will use $\hat{m}_S(x|t)$ in place of $m(x|t)$ in the following for numerically derived solutions. Fortunately, the approximation error resulting from this truncation can be made arbitrarily small. For a detailed discussion, we refer the reader to the previous Chapter 4.2.2 and to Hasenauer et al. (2012a).

The composition of the overall label distribution, $m(x|t)$, from the label intensities of individual cells, as well as subpopulations of distinct cell types j and division numbers i, and their change over time, are illustrated in Figure 5.1. This illustration also elucidates the relationship between the information contained in the model and the labeling experiments, or more precisely, how the model information is related to the processes in such proliferation experiments as well as to their measurement output.

5.3 Example: CFSE-labeled stem cell population

To illustrate the capability of our model, we will now investigate an example of a cell population with asymmetric cell division and hence naturally more than one cell type. We will show how this can be reflected by the developed model, and how our model serves to elucidate the proliferation properties.

5.3.1 Cell population with asymmetric cell division

For the example studied here, we consider a cell population which descends from a stem cell pool, but where cells can commit towards a second, more mature stage upon cell division. In asymmetric cell division, one daughter cell is again a stem cell, while the second cell is a committed cell. In symmetric cell division in contrast, both daughter cells are of the same cell type (either stem cells or committed cells) (Morrison and Kimble, 2006). The role of asymmetric cell divisions is of high relevance in understanding the maintenance or homeostasis of (stem) cell populations, and is thus the subject of numerous publications (see for example, Knoblich (2008); Morrison and Kimble (2006); Roegiers and Jan (2004), and references therein).

For this case study, we investigate an exemplary CFSE-labeled multi-cell type population for which we make the following assumptions:

- There are two distinct cell types: stem cells ($j = 1$) and committed cells ($j = 2$).

- Only stem cells divide ($\forall i : \alpha_i^1 \geq 0$), but committed cells do not: $\forall i : \alpha_i^2 = 0$.

- Cell type transitions only occur in the context of asymmetric cell division, but not without cell division: $\forall i, j_1, j_2 : \delta_i^{j_2 j_1} = 0$.

The dependency of parameters on time is omitted, $\beta_i^j(t) = \beta_i^j > 0$, $\alpha_i^j(t) = \alpha_i^j \geq 0$, and $w_i^{11}(t) = w_i^{11}, w_i^{21}(t) = w_i^{21} \in [0,1]$, with $w_i^{21} = 1 - w_i^{11}$. The resulting model structure for this scenario is illustrated in Fig. 5.2, with subpopulations (i, j) depicted as compartments, and fluxes due to cell division and cell type transitions.

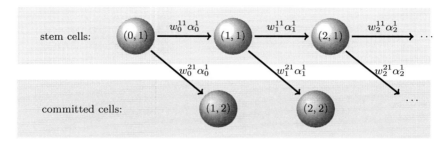

Figure 5.2: Model structure for the example of asymmetrically dividing stem cells. Illustration of cell types (rows) and division numbers (columns): The circles denote the subpopulations (i, j) of cells with division number i and cell type j, and the arrows denote the fluxes between these subpopulations.

With this, the resulting model equations are, $\forall i \in \mathbb{N}_0$:

$$\frac{\partial n(x,i,1|t)}{\partial t} - k(t)\frac{\partial\,(x\,n(x,i,1|t))}{\partial t} = -\beta_i^1 n(x,i,1|t) - \alpha_i^1 n(x,i,1|t)$$
$$+ \begin{cases} 0 & ,i = 0 \\ 2\alpha_{i-1}^1 w_{i-1}^{11} n(x,i-1,1|t) & ,i \geq 1 \end{cases} \quad (5.19)$$
$$\frac{\partial n(x,i,2|t)}{\partial t} - k(t)\frac{\partial\,(x\,n(x,i,2|t))}{\partial t} = -\beta_i^2 n(x,i,2|t) + 2\alpha_{i-1}^1 w_{i-1}^{21} n(x,i-1,1|t)$$

with initial conditions

$$n(x,0,1|0) = n_{\text{ini}}(x),\ \forall (i,j) \neq (0,1) : n(x,i,j|0) = 0. \quad (5.20)$$

Thanks to the decomposition (5.9) which can be conducted for this model as outlined above, the model can be analyzed in two independent parts: The cell population dynamics are determined by a system of ODEs, and the label dynamics are given by a set of linear PDEs.

5.3.2 Analysis of cell population dynamics

First, let us analyze the proliferation properties and the resulting cell population dynamics, in particular the subpopulations of the two distinct cell types. The dynamics of the subpopulation sizes $N(i,j|t)$ can be studied via the emerging DC model, $\forall i \in \mathbb{N}_0$:

$$\frac{dN(i,1|t)}{dt} = -\beta_i^1 N(i,1|t) - \alpha_i^1 N(i,1|t)$$
$$+ \begin{cases} 0 & ,i = 0 \\ 2\alpha_{i-1}^1 w_{i-1}^{11} N(i-1,1|t) & ,i \geq 1 \end{cases} \quad (5.21)$$
$$\frac{dN(i,2|t)}{dt} = -\beta_i^2 N(i,2|t) + 2\alpha_{i-1}^1 w_{i-1}^{21} N(i-1,1|t)$$

with initial conditions

$$N(0,1|0) = N_{\text{ini}},\ \forall (i,j) \neq (0,1) : N(i,j|0) = 0. \quad (5.22)$$

An analytical solution for this model is presented in Appendix H. Let us now focus on the case where parameters depend on cell types, but not on division numbers, $\forall i : \beta_i^j = \beta^j > 0$, $\alpha_i^j = \alpha^j \geq 0$, and $w_i^{11} = w^{11}, w_i^{21} = w^{21} \in [0,1]$, with $w^{21} = 1 - w^{11}$. For this model the analytical solution is

$$
\begin{aligned}
N(i,1|t) &= \frac{(2w^{11}\alpha^1 t)^i}{i!} e^{-(\beta^1 + \alpha^1)t} N_{\text{ini}}, \\
N(i,2|t) &= \frac{(2w^{11}\alpha^1)^i \frac{w^{21}}{w^{11}}}{(i-1)!} e^{-\beta^2 t} \cdot \left(\int_0^t \tau^{i-1} e^{(\beta^2 - \beta^1 - \alpha^1)\tau} d\tau \right) N_{\text{ini}}.
\end{aligned}
\tag{5.23}
$$

The derivation of this solution (5.23) can be found in Appendix J.

We will now turn towards important questions in the context of proliferating stem cell populations, and tackle them by exploiting the derived model. Since high-quality stem cell pools are rare and costly to achieve, it is relevant to identify conditions under which they will not diminish too fast. Therefore, a crucial question in the introduced scenario of stem cell division is:

(Q1) How can the total amount of stem cells be kept constant, and which proliferation parameters have to be controlled in order to achieve this?

In order to answer this question, we consider the total amount of stem cells

$$
\bar{N}(1|t) = \sum_{i=0}^{\infty} N(i,1|t) = e^{((2w^{11}-1)\alpha^1 - \beta^1)t} N_{\text{ini}},
\tag{5.24}
$$

for which the derivation is provided in Appendix K.

The total number of stem cells is determined by the sign of the exponent in (5.24) as follows:

case (A): $\bar{N}(1|t)$ increases $\Leftrightarrow \beta^1 < (2w^{11} - 1)\alpha^1$,

case (B): $\bar{N}(1|t)$ is constant $\Leftrightarrow \beta^1 = (2w^{11} - 1)\alpha^1$, or

case (C): $\bar{N}(1|t)$ decreases $\Leftrightarrow \beta^1 > (2w^{11} - 1)\alpha^1$.

The parameter subspace on which the amount of stem cells remains constant, case (B), along with illustrative projections, is depicted in Fig. 5.3. The depicted surface separates the parameter subspace (above) for which the amount of stem cells increases, case (A), from the parameter subspace (below) where it decreases, case (C).

Restricting to biologically feasible parameter values $\alpha \in \mathbb{R}_+$ and $\beta \in \mathbb{R}_{++}$, the model reveals that a necessary condition to conserve the amount of stem cells is $w^{11} > 0.5$: The probability that a daughter cell is a stem cell must be higher than the probability for being a committed cell. This is an important finding: On the one hand, it will be highly relevant to identify factors, such as biochemical components and environmental circumstances, that determine the cell type commitment of dividing cells. On the other hand, if the probability of cell type transition upon division is too high and cannot be influenced, then it will be impossible to conserve the stem cell pool in the long term. These observations, along with the feasible parameter value ranges, provides an answer to question (Q1).

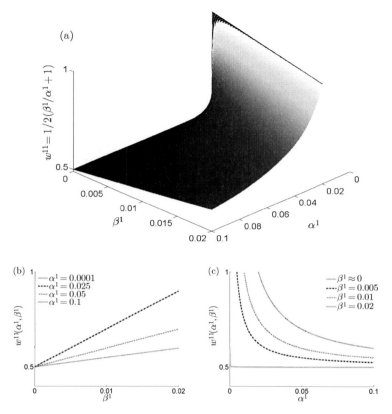

Figure 5.3: (a) Parameter subspace on which the total amount of stem cells remains constant, case (B). For parameter values above the plane, $\bar{N}(1|t)$ increases (case (A)), whereas below the plane, $\bar{N}(1|t)$ decreases (case (C)). (b,c) Projections for exemplary values in the β^1-w^{11}-plane and in the α^1-w^{11}-plane.

A second interesting question originates from two conflicting goals: Minimization of the number of committed cells, in order to keep the stem cell pool as "pure" as possible; or maximization of the number of committed cells, to build up or repair tissue. One may therefore pose the following question:

(Q2) What is the expected ratio of the number of committed cells to the number of stem cells, in dependence of the proliferation parameters?

As also demonstrated, for example, by Balázsi et al. (2011), this question of optimizing the ratio of committed-to-stem cells may be addressed by mathematical modeling to gain additional biological insight. To approach this question, we consider the committed-to-stem cells ratio, for $\beta^2 + (2w^{11} - 1)\alpha^1 - \beta^1 \neq 0$ given by

$$\frac{\bar{N}(2|t)}{\bar{N}(1|t)} = \frac{2w^{21}\alpha^1}{\beta^2 + (2w^{11} - 1)\alpha^1 - \beta^1} \left(1 - e^{-(\beta^2 + (2w^{11} - 1)\alpha^1 - \beta^1)t}\right). \tag{5.25}$$

For a detailed derivation, the reader is referred to Appendix L. Apparently, the sign of the denominator in (5.25) determines the dynamics of the committed-to-stem cells ratio, as follows:

case (i): If $\beta^1 - \beta^2 < (2w^{11} - 1)\alpha^1$, then the committed-to-stem cells ratio $\bar{N}(2|t)/\bar{N}(1|t)$ will grow with saturation to the limit

$$R_{2/1}^{\max} := \lim_{t \to \infty} \frac{\bar{N}(2|t)}{\bar{N}(1|t)} = \frac{2w^{21}\alpha^1}{\beta^2 + (2w^{11} - 1)\alpha^1 - \beta^1}. \tag{5.26}$$

case (ii): If $\beta^1 - \beta^2 = (2w^{11} - 1)\alpha^1$, then the committed-to-stem cells ratio is linear

$$\frac{\bar{N}(2|t)}{\bar{N}(1|t)} = 2w^{21}\alpha^1 t. \tag{5.27}$$

case (iii): If $\beta^1 - \beta^2 > (2w^{11} - 1)\alpha^1$, then the committed-to-stem cells ratio $\bar{N}(2|t)/\bar{N}(1|t)$ will approach exponential growth, thus

$$\lim_{t \to \infty} \frac{\bar{N}(2|t)}{\bar{N}(1|t)} = \infty. \tag{5.28}$$

Further effects of the individual parameters on the maximum ratio can be determined from analyzing the respective derivatives $\partial R_{2/1}^{\max}/\partial \cdot$:

- Increasing the stem cell death rate, β^1, always reduces the number of stem cells and thereby increases the proportion of committed cells.

- Similarly, increasing the probability for a newborn cell to be of the committed cell type, w^{21}, always increases the maximum fraction of committed cells.

- Conversely, increasing the death rate of committed cells, β^2, or the probability that a newborn cell is again a stem cell, w^{11}, both reduce the limiting value $R_{2/1}^{\max}$.

- In contrast, the effect of the cell division rate α^1 is less obvious, but interestingly depends on the relationship of death rates: If the death rate of committed cells is higher than the death rate of stem cells, $\beta^2 > \beta^1$, increasing α^1 has an increasing effect on the maximum fraction of committed cells, $R_{2/1}^{\max}$, meaning it yields a higher proportion of committed cells. Contrary, for $\beta^2 < \beta^1$, increasing the cell division rate α^1 has a reducing effect, meaning it yields a higher proportion of stem cells.

This model-based analysis of the determinants for the ratio of committed cells to stem cells provides answers to question (Q2).

With this analysis of population dynamics, we have exemplified how the presented model can serve to investigate the relationship between proliferation parameters and proliferation dynamics in a quantitative way, already by analyzing solely the ODE part of the model without considering the label dynamics. Next, we will also look at the CFSE label dynamics and connect them with the population dynamics, in order to provide a link between these findings and the measurement output from CFSE labeling experiments.

5.3.3 Analysis of label distribution dynamics

For inspecting the label dynamics, let us first consider an exemplary simulation of scenario (B), that is a constant stem cell population $\overline{N}(1|t)$ and a limited committed-to-stem cells ratio, $\overline{N}(2|t)/\overline{N}(1|t)$. The parameters for this simulation were chosen as follows: $\alpha^1 = 0.5[1/d]$, $\beta^1 = \beta^2 = 0.025[1/d]$, $w^{11} = 0.525$, $w^{21} = 0.475$, and labeling parameters $k(t) = 0.07[1/d]$, $\gamma = 2$. For the solution of the ODE system (5.10), a truncated system was used as discussed in the previous chapter in Section 4.2.2, with $S = 20$ division numbers. Besides being biologically reasonable, this choice of considering 20 divisions also ensured a sufficiently small truncation error, as pointed out below. The simulation of this model took less than 0.5 seconds on a standard laptop.

The results of this scenario for day $t = 4[d]$ are depicted in Figure 5.4 on subsequent steps of detail that also illustrate the assemblage of the model solution from the individual parts: First, the probability densities $p(x|i,t)$ are derived as the PDE solution (5.15), depending on time and division number. With increasing division number, the respective probability density is located more to the left, corresponding to lower label concentration, as seen in the depicted division numbers $i = 0, \ldots, 5$. Similarly, for increasing time, the probability densities move to the left (not shown), due to the degradation of label with degradation rate $k(t)$.

Next, these probability densities, $p(x|i,t)$, are weighted by the respective number of cells of each subpopulation, $N(i,j|t)$, in order to derive the individual number densities $n(x,i,j|t) = N(i,j|t)p(x|i,t)$, by (5.9). The number densities of stem cells and committed cells are for the same division number located at the same values of label concentration x, as the location is determined by the probability densities, $p(x|i,t)$ and the label dynamics are independent of the cell type. However, the number densities differ in height, because stem cells and committed cells have different numbers of cells, $N(i,j|t)$. Finally, by summing over all division numbers $i = 0, \ldots, S-1$ and all cell types $j = 1, 2$, the overall label distribution, $m(x|t)$, is obtained. The last panel of Figure 5.4 shows $m(x|t) \cdot x$, instead of the overall label distribution $m(x|t)$. In this way we account for the increasing bin width from the logarithmic x scale, similar to the output from measurements which is given by cell counts of intensity channels.

As mentioned above, the model was solved using a truncated system of ODEs (5.21), for $i = 0, \ldots, S-1$ with $S = 20$. While for the considered example analytical solutions of the ODE system are available, (5.23), the numerical solution was found to yield even more precise simulation results, probably due to numerical instabilities in evaluating the analytical expressions, which involve factorials and exponents in i. The truncation error is upper bounded, as pointed out in the previous Chapter 4 and Appendix C, and this upper bound (C.24) can be computed using the analytical solution (C.25). The upper bound for the truncation error in this case is, at time point $t = 5$,

$$\frac{||m(x|t) - \hat{m}_S(x|t)||}{||m(x|0)||} \leq 1.22 \cdot 10^{-4}, \tag{5.29}$$

while for earlier time points $t \in (0,5)$ it will be even lower. This exemplifies that the truncation with $S = 20$ is of very good accuracy, taking into account that the original model was given by a system of coupled PDEs (5.19).

After having discussed in detail how the overall label distribution $m(x|t)$ is obtained efficiently by means of scenario (B), let us now survey the differences in this overall label distribution $m(x|t)$ between distinct scenarios. Whereas for the dynamics of cell numbers, $N(i,j|t)$, this may seem obvious and was discussed in Subsection 5.3.2, it may seem less intuitive for the overall label distribution, $m(x|t)$, since it is assembled from weighted probability densities, as we have outlined just above.

The fact that distinct scenarios of proliferation behavior indeed yield distinct dynamics in the overall label distribution is demonstrated in Figure 5.5. We show representative simulation results from three different scenarios of case (A), (B), and (C), respectively. Of the three possibilities for case (C), that is case (i), (ii), or (iii), an exemplary scenario was chosen of case (iii) and will be denoted by (Ciii). The chosen exemplary parameter values were, for all scenarios, $\alpha^1 = 0.5$, $\beta^1 = \beta^2 = 0.025$, and furthermore for scenario (A) $w^{11} = 0.7$, $w^{21} = 0.3$, for scenario (B) $w^{11} = 0.525$, $w^{21} = 0.475$, and for scenario (Ciii) $w^{11} = 0.3$, $w^{21} = 0.7$. According to the analytical results obtained in Subsection 5.3.2, these three scenarios will yield qualitatively different dynamical behavior. Simulation results are depicted for the overall label distribution, $m(x|t)$, at three different time points, $t = 1, 3, 5[d]$, and for the number of cells over a time span $t \in [0, 5]$, for both cell types $j = 1, 2$ and the first six division numbers $i = 0, \ldots, 5$.

For scenario (A), with increasing time the distribution $m(x|t)$ gets quite broad, in terms of exhibiting many peaks, corresponding to cells in the population with very different division numbers. This effect is less pronounced for scenario (B), whereas for scenario (Ciii) the distribution is even more narrowed, featuring only a few pronounced peaks. The plots to the right hand side show the according cell population dynamics for each scenario. These dynamics of the cell numbers, $N(i, j|t)$, accomplish the noticed differences between the scenarios: Whereas in scenario (A) the depicted subpopulations are of roughly the same magnitude, this is less the case for scenario (B), and for scenario (C) the differences between cell numbers of different subpopulations are even more pronounced.

In view of the differences between the scenarios, the model seems promising to distinguish between cases of qualitatively different dynamics, if solely given the overall label distribution. If such a model is used for parameter estimation, it can exploit the full information from the data to identify proliferation properties. The example of a CFSE-labeled stem cell population investigated here illustrates how crucial differences in population dynamics, such as whether the amount of stem cells is increasing or decreasing, exhibit noticably different label distributions.

Individual label densities:

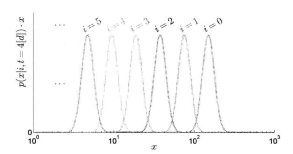

Individual number densities:

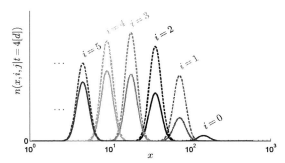

Overall label distribution:

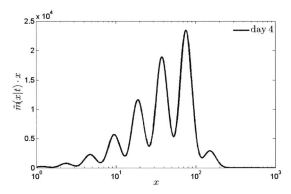

Figure 5.4: Assembling the overall label distribution in the cell population: From the individual label densities, $p(x|i,t)$, and the numbers of cells, $N(i,j|t)$, the individual number densities, $n(x,i,j|t)$, are assembled. The number densities are depicted separately for the two cell types: stem cells (solid lines –) and committed cells (dotted lines - -). Summing over all cell types j and division numbers i yields the overall label distribution, $m(x|t)$. The simulation results are for scenario (B) (constant number of stem cells) at time point $t = 4[d]$, parameters $\alpha^1 = 0.5[1/d]$, $\beta^1 = \beta^2 = 0.025[1/d]$, $w^{11} = 0.525$, $w^{21} = 0.475$, $k(t) = 0.07[1/d]$, $\gamma = 2$, and are only shown for the first six division numbers $i = 0, \ldots, 5$.

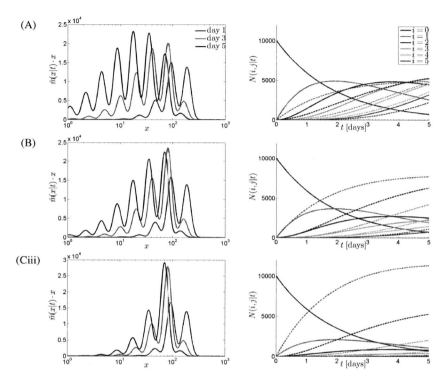

Figure 5.5: Simulation results for three distinct scenarios, for exemplary parameter values. Parameters were for all scenarios $\alpha^1 = 0.5$, $\beta^1 = \beta^2 = 0.025$, and depending on the respective scenario (A) $w^{11} = 0.7$, $w^{21} = 0.3$, (B) $w^{11} = 0.525$, $w^{21} = 0.475$, (Ciii) $w^{11} = 0.3$, $w^{21} = 0.7$. Left hand side: The overall label distribution, $m(x|t)$, at three time points, $t = 1[d]$ (rightmost line), $t = 3[d]$, $t = 5[d]$ (leftmost line). Right hand side: The numbers of cells, $N(i,j|t)$, for the two cell types: stem cells (solid lines –) and committed cells (dotted lines - -), for the first six division numbers $i = 0$ (line starting topmost and line equal zero), ..., $i = 5$ (least increasing lines).

5.4 Summary and discussion

In this chapter, we presented a model for CFSE-labeled multi-cell type populations, and thereby provide a solution to Problem 5.1. The proposed model allows to represent the dynamics of proliferating cell populations which comprise multiple cell types, while also accounting for division numbers and CFSE label dynamics. To the best of our knowledge, this is the first model for CFSE-labeled populations that is structured by cell types, division numbers, and label concentration at the same time.

Incorporating the specific properties of CFSE labeling into the general model class of Chapter 4, we obtained a decomposable model, which is a powerful result. The model equations, originally given by a coupled set of PDEs, are split into a system of ODEs and a set of decoupled linear PDEs. Thereby, the model becomes amenable to analysis and solution approaches for decomposable population models, as they have been presented in Chapter 4. The modeling and analysis tools that have been developed on a general level therein could now be readily employed for the CFSE-specific model.

Compared to other population models that account for cell types and a continuous-valued property, but importantly no division numbers (see, for example, Gyllenberg and Webb (1990)), our modeling approach tremendously reduces the computational burden. Regarding the CFSE-specific labeling dynamics in turn, there exist other population models for CFSE which are division- and label-structured (Hasenauer et al., 2012a; Schittler et al., 2011; Thompson, 2011). These models however come with the considerable drawback that they do not account for cell types, whereas their solution is of equal computational complexity (Hasenauer et al., 2012a; Schittler et al., 2011) as for the here presented model.

We demonstrated how our model can serve to analyze proliferation dynamics in multi-cell type populations by studying the example of stem cells, which may partly differentiate into a committed cell type upon asymmetric or symmetric cell division. As we have shown, the analysis of proliferation properties based on our model can reveal parameter settings for which qualitatively distinct population dynamics arise. This renders the model especially valuable if biological experiments are time consuming or not at hand, or it can be used to complement experimental investigations, for example by comparing and validating hypotheses. In addition, we showed how the resulting measurements of the overall label distributions differ between scenarios with qualitatively different dynamics. Our model may thus elucidate cell type-specific proliferation properties, even if they are not directly visible in the measurement output.

Our model can be employed to infer proliferation parameters from CFSE data, as done already with existing models (Banks et al., 2010, 2011; Luzyanina et al., 2007a,b; Thompson, 2011; Hasenauer, 2012). Therefore, the overall label density had to be compared to measurement data, which was out of the scope of this work. In this way, our model offers the possibility of studying the subpopulation structure using typical measurements from CFSE labeling experiments, which so far was not possible with existing models. The model developed here may as well extend to other labeling techniques that resemble CFSE in their properties (Quah et al., 2007). For cell populations containing cell types with distinct proliferation properties, our model promises to allow for more realistic dynamics and a better explanation of data from cell proliferation assays.

Since parameter estimation from data was not covered in this chapter, we also neglected the autofluorescence for the model output. However, the successful incorporation of the autofluorescence into closely related models suggests that similar approaches apply to the model presented here. Such an approach is pursued for example in the next chapter on modeling BrdU-labeled cell populations. A comprehensive analysis and discussion of incorporating the autofluorescence into models of CFSE-labeled cell populations is furthermore provided by Hasenauer et al. (2012a).

Chapter 6

Modeling of BrdU-labeled multi-cell type populations

In this chapter, we develop a specific model for multi-cell type populations labeled with Bromodeoxyuridine (BrdU). Exploiting the general model class that was presented in Chapter 4 for proliferating multi-cell type populations in labeling experiments, we account for the specific properties of BrdU labeling, from which we successively develop a suitable model. Section 6.1 provides the necessary background on BrdU labeling and available models, after which we formulate the central problems addressed in this chapter. Section 6.2 is dedicated to the comprehensive development of the model, along with a mathematically rigorous model for the DNA label segregation process, an efficient solution approach, and the connection of model output and measurement data. In Section 6.3, we exploit the derived model for two simulation studies: First, we investigate the specific characteristics observed in BrdU data and show how some potential sources for these observations can be excluded with the use of our model. Second, we employ our model to estimate proliferation parameters from BrdU data, and compare the results and performance to existing approaches. The chapter concludes with a short summary and discussion in Section 6.4.

Parts of the work here are based on Schittler et al. (2013a) and Schittler et al. (in prep.).

6.1 Background and problem formulation

6.1.1 BrdU labeling experiments

A common labeling technique for studying cell proliferation employs Bromodeoxyuridine (BrdU) as a so-called DNA label[1]. Several decades ago, BrdU has been discovered as a synthetic analog for thymidine (Gratzner, 1982), one of the four nucleotides building up the DNA (Alberts et al., 2000). Upon DNA replication in an environment that contains BrdU, the label is taken up by the dividing cells and incorporated during mitosis into the newly synthesized DNA. Once incorporated, it remains in the DNA, thus there is no degradation of the BrdU label. If a cell divides that already contains BrdU in its DNA, it is passed on to the daughter cells since half of each newly born cell's DNA will come from the mother cell (Kee

---

[1]Although DNA labels are rather integrated into the DNA and thus the terms "DNA label" or "DNA labeling" may be misleading, they have been established as the common terminology.

et al., 2002). The BrdU contained in a cell can then be detected using antibodies and flow cytometry (Gratzner, 1982).

There has been a discussion on whether or to what extent BrdU affects cellular functions such as proliferation and differentiation (see, for example, Hill et al. (1974); Lehner et al. (2011), and references therein). If administered over extended time periods or at high dosages, it causes mutations and cells exhibit a mitogenic effect (Kiel et al., 2007; Lehner et al., 2011; Hoshino et al., 1985; Takizawa et al., 2011; Wilson et al., 2008). Because of these objections, alternative labels have been suggested such as deuterated water, but which in contrast requires detection by mass spectrometry, making it more complicated and expensive than BrdU labeling.

BrdU has been applied in many proliferation studies on diverse cell systems, such as neural cells (Lehner et al., 2011), tumor cells (Hoshino et al., 1985), haematopoietic stem cells (Kiel et al., 2007; Takizawa et al., 2011), and lymphocytes (De Boer and Perelson, 2013; Mohri et al., 1998; Tough and Sprent, 1998), to give just some examples. A useful property of BrdU is that it may be combined with other labeling markers, such as CFSE to have an additional proliferation marker, or green fluorescence protein to track cell lineages (Takizawa et al., 2011; Wilson et al., 2008). Recently, labels have been established that are similar to BrdU as they also act as thymidine analogs, such that analytical and modeling approaches developed for BrdU may extend to these labeling techniques (Conboy et al., 2007; Mull and Asakura, 2012, and references therein).

6.1.2 Available models for BrdU-labeled cell populations

To obtain data from BrdU labeling experiments, cells are sent through a flow cytometer and for each single cell it is recorded into which intensity bin it falls. Thus, data from BrdU labeling experiments generally come as histograms (binned snapshot data) from a sample of label intensities $\{l^\nu\}, \nu = 1, \dots, M$, of M measured cells. The obtained BrdU intensities are commonly spread out over a wide range of values, and the data do not allow for a straightforward quantitative mapping from an individual cell's BrdU intensity to its number of undergone divisions. This is in contrast to data from CFSE labeling experiments, which often exhibit clearly separated peaks, each corresponding to a distinct division number (confer the previous chapter).

For this reason, BrdU data is analyzed via choosing a certain threshold l_θ, which is usually regarded as the threshold above the autofluorescence. Importantly, the autofluorescence in BrdU measurements has been shown to differ over a broad range between cells (Takizawa et al., 2011), thus the threshold is often not clearly determined and highly dependent on the sample of measured cells. Various threshold choices have been suggested (Kiel et al., 2007; Mohri et al., 1998; Parretta et al., 2008; Wilson et al., 2008), further emphasizing the variability of data interpretation between different studies. Based on the chosen threshold, the number of BrdU positive cells is defined by $|\{l^\nu | l^\nu \geq l_\theta\}|$, and the number of BrdU negative cells by $|\{l^\nu | l^\nu < l_\theta\}|$. In this way, the data set is boiled down into basically two numbers for each measured time point. Unfortunately, much of the information contained in the original data is thereby lost, and a heavy reliance on the threshold value is introduced. The reason for this data processing is probably that the original data do not promote a straightforward interpretation, but require mathematical modeling. Most mathematical models up to date are rather simple as they consider solely the number of BrdU positive and negative cells (Bonhoeffer et al., 2000;

De Boer et al., 2003b,c; Kiel et al., 2007; Mohri et al., 1998; Glauche et al., 2009; Parretta et al., 2008; Wilson et al., 2008). For a more sophisticated interpretation of BrdU data, accordingly sophisticated mathematical models are inevitable.

A step towards exploiting more of the information provided in BrdU data is offered by models that incorporate the mean fluorescence intensity or the mean BrdU content (Bonhoeffer et al., 2000; Ganusov and De Boer, 2013). Still, the information that is originally contained in the data as representing a distribution ("intensity profile") can not be fully used by these models. It was suggested by De Boer and Perelson (2013) that, for extracting more information from BrdU data, a new model class should account for full intensity profiles. A related discrepancy of common available mathematical models for BrdU labeling is the observation that individual cells may not have picked up the same fixed value of label content, although having undergone the same number of divisions (Takizawa et al., 2011). To overcome this drawback, several models introduced a factor that should reflect the population-wide proportion of label uptake upon division, which was also considered to be possibly time-dependent (Asquith et al., 2002; Bertuzzi and Gandolfi, 2000; Bonhoeffer et al., 2000; Debacq et al., 2002; Glauche et al., 2009). In the following, we will refer to the effective label uptake into cells upon divisions as the "labeling efficacy".

To the best of the author's knowledge, the first model class which accounts for the full intensity profile, and also incorporates a general labeling efficacy distribution, was presented by Schittler et al. (2013a). However, this model class still exhibits several deficits: First, it does not consider cell types with distinct proliferation properties. In this way, the model class is not suited for many cell populations which are known or suspected to comprise multiple cell types. Second, the labeling efficacy was rather modeled in a phenomenological way, without a deeper consideration of the unterlying processes of label uptake into and label inheritance via chromosomes. Third, for a direct connection to BrdU data, one would need to incorporate the autofluorescence, which was neglected in this previous work of Schittler et al. (2013a).

Since the chromosomes are the carriers of DNA labels, the inheritance of label from a dividing mother cell to its daughter cells can be expected to depend in some way on the segregation of chromosomes. As will also be discussed in detail later, the process of DNA label segregation in general can not be assumed to lead to equal partitioning, that is halving of the label, to the daughter cells. Nevertheless, most available models assume halving of the label upon division. Only few models have introduced heuristic approaches that allow for a more general segregation factor, given by a random variable drawn from a probability distribution (Glauche et al., 2009). But notably, so far it has not been investigated in a rigorous mathematical framework whether the dependency of DNA labels on chromosome segregation can be neglected, and if not so, how the label segregation could be modeled suitably.

To comply with the need to account for multiple cell types, numerous models have been introduced for BrdU labeling that incorporate distinct cell types (Bonhoeffer et al., 2000; De Boer et al., 2003b; Ganusov and De Boer, 2013; Glauche et al., 2009; Grossman et al., 1999; Parretta et al., 2008; Ribeiro et al., 2002; Wilson et al., 2008), mostly accompanied by restrictive assumptions such as not more than two distinct cell types, or/and steady state assumptions on each cell type subpopulation. However, most of them in turn do not consider the label intensity information from data, but solely distinguish between BrdU positive and negative cells. Only some of these models consider, as mentioned above, the mean fluorescence intensity or the mean BrdU content in order to provide a connection to the measured

BrdU intensities (Bonhoeffer et al., 2000; Ganusov and De Boer, 2013), but none of these cell type-structured models for BrdU labeling provides the full intensity profiles. Especially when distinct cell types are present in a population, thus increase the number of proliferation parameters, it becomes increasingly challenging to identify these multiple parameters. The more it is important to exploit all available information from BrdU data, and thus to develop models that allow to construct the full intensity profiles.

6.1.3 Problem formulation

As has been outlined, there is a lack of models for BrdU labeling that account for cell types, division numbers, and at the same time can exploit the full intensity distribution. The rigorous development of such a model requires a mathematical description of the label uptake into cells, as well as for the label inheritance from mother to daughter cells. The inheritance of DNA labels, such as BrdU, is tightly connected to the segregation of chromosomes from mother to daughter cells. This renders the mathematical modeling of the label uptake and label inheritance nontrivial. Former models neglected the detailed mechanisms of chromosome segregation, but instead used phenomenological approximations for the combined effect of label uptake and label inheritance, such as the assumption that each daughter cell inherits exactly half of the label (Ganusov and De Boer, 2013; Schittler et al., 2013a). Because, to the best of our knowledge, up to date no mechanistic models for DNA label segregation are available, it is also unclear whether or to what extent such phenomenological approaches are valid.

To remedy this deficit, we here aim to derive a mathematically rigorous description of this process. It is thereby important that the derived mathematical description allows for an efficient evaluation. This challenge is captured by the following problem that we address:

Problem 6.1 *(Modeling of DNA label segregation) Given the division number of a mother cell, and the labeling efficacy in the environment, develop a model for the probability distribution of DNA label content of the daughter cells.*

As it will be seen, the mathematical description of DNA label segregation is nontrivial, and subtle modeling approaches need to be elaborated in order to obtain a computationally tractable model. If the DNA label segregation would be captured by a rigorous mathematical model, one could proceed to derive a model for BrdU-labeled multi-cell type populations that reproduces the full label intensity distribution.

For reconstruction of the full label intensity distribution, no rigorously derived model is available at all up to date. The explicit modeling of cell types is essential since many of the cell populations investigated in BrdU labeling studies comprise multiple subpopulations with distinct proliferation properties, whereas division numbers need to be modeled in order to achieve the proliferation parameters of interest. Modeling the distribution of BrdU label intensity, in turn, would enable to exploit the full information from data, which is especially necessary if proliferation parameters of multiple cell types should be identified from measurements. Since it is vital to make such models available, we address the following problem:

Problem 6.2 *(Modeling of BrdU-labeled proliferating multi-cell type populations) Given the general model for a multi-cell type, proliferating, labeled cell population, and the labeling properties of BrdU, develop an efficient multi-cell type population model for BrdU labeling.*

Once again, we can employ the general DsCL model class that was developed in Chapter 4 to specify a model for BrdU-labeled proliferating multi-cell type populations. The specific properties of BrdU labeling can be exploited to develop a model which allows for an efficient solution scheme.

6.2 Specific model for BrdU-labeled multi-cell type populations

In this section we will specify the model for BrdU labeling, whereas a crucial part will concern the segregation of DNA label from mother to daughter cells. Following the model development, we will propose an efficient approach to solve the model.

6.2.1 Model specification

We will now subsume the properties of BrdU labeling in mathematical terms, which can later be substituted into the DsCL model equations (4.1), in order to derive a specific model for BrdU.

As mentioned above, BrdU is incorporated into the DNA as an analog for thymidine, and thus not degraded (Gratzner, 1982). Therefore, the decay term is simply $\forall x, t$:

$$\nu(x, t) = 0. \tag{6.1}$$

A BrdU labeling experiment typically comprises two time intervals with distinct labeling conditions, $K = 2$, namely an uplabeling phase (sometimes also called the "pulse period"), $(0, T_1]$, and a delabeling phase (sometimes also called the "chase period"), $(T_1, T_2]$. The division sequence will consequently be represented by a vector $\mathbf{i} = (i_1, i_2)$. At the beginning of a BrdU experiment, no label is contained in the cells, independent of their cell type. Furthermore, it can be assumed that the label incorporation is independent of the cell type. Together, this can be reflected by the initial number density being nonzero only at $x = 0$ in the yet undivided cells, $\forall j$:

$$\begin{aligned} \mathbf{i} = 0^T : \ & n(x, 0^T, j|0) = n_{\text{ini},j}(x) = N_{\text{ini},j} \cdot \delta(x) \\ \forall \mathbf{i} \neq 0^T : \ & n(x, \mathbf{i}, j|0) = 0. \end{aligned} \tag{6.2}$$

During the uplabeling phase, BrdU is administered and gets incorporated with a label uptake which is invariant from the beginning of the experiment until the end of the uplabeling phase,

$$\forall t \in [0, T_1] : \ p_{\text{lab}}(x|t) = p_{\text{up}}(x). \tag{6.3}$$

During the delabeling phase, BrdU is withdrawn and thus can not be incorporated any more,

$$\forall t \in (T_1, T_2] : \ p_{\text{lab}}(x|t) = \delta(x). \tag{6.4}$$

The processes of label uptake during uplabeling and label inheritance require a more comprehensive description, as illustrated exemplarily in Figure 6.1 and subsumed generally in the following. Let a cell have n_{chrom} chromosomes, which each contains a segment of (double-stranded) DNA helix. This means that a cell has in total $2n_{\text{chrom}}$ strands. In the example of Figure 6.1, cells have $n_{\text{chrom}} = 5$ chromosomes, thus the cell at the beginning has $2n_{\text{chrom}} = 10$ strands, of which in this example 6 are labeled (red).

(P1) Upon mitosis, each chromosome is duplicated by DNA synthetization. If at the time of DNA synthetization the environment provides a thymidine analog (such as BrdU), then a certain proportion of the thymidine is substituted by the analog, thus it is incorporated as a label into the newly synthesized DNA.

(P2) The amount of effectively uptaken label may fluctuate between division events, but can be assumed to be equal for all strands created in the same division event. Furthermore, since a DNA strand contains ten thousands to hundred thousands of thymidines, it can be assumed that the proportion of thymidine analog is the same for each strand synthesized in the same division event.

(P3) After chromosomes have been duplicated, the cell contains $2n_{\text{chrom}}$ DNA helixes, thus $4n_{\text{chrom}}$ strands. Each DNA helix segment now consists of an "old" strand, which originates from the mother cell, and a "new" strand, which has just been synthesized.

(P4) Upon cell division, one mother cell divides into two daughter cells.

(P5) For each chromosome, existent in two copies, daughter cell 1 receives one randomly chosen copy, and daughter cell 2 receives the other copy. Which daughter cell receives which of the two copies of a chromosome is random, and independent between individual chromosomes.

Afterwards, each daughter cell possesses one copy of each chromosome. As a result of the segregation process, all combinations of old strands (having received copy 1 or 2 for the 1st chromosome,..., copy 1 or 2 for the n_{chrom}-th chromosome) are possible, and if the old strands differ in their proportion of uptaken label, these combinations yield different inherited label concentration. In contrast, the n_{chrom} new strands equal each other in their proportion of uptaken label, thus in their label concentration. In the exemplary illustration in Figure 6.1, the two daughter cells end up with one having 9 labeled strands, and the other having 7 labeled strands, while this difference between the two cells purely results from the randomness in (P5). Each of the two daughter cells possesses 5 strands that have been labeled in this recent division event (marked in blue). In addition, daughter cell 1 has 4 previously labeled strands (marked in red), of which three originate from the same division event and thus have the same value of uptaken label (horizontal lines), whereas one originates from a distinct division event and thus has a different value of uptaken label (diagonal lines). Daughter cell 2 has received two strands that have been synthesized in the same previous division event, thus have the same uptaken label (horizontal lines).

This shows that the whole process is to some extent deterministic, namely because a) each daughter cell receives one copy of each chromosome, and b) each chromosome contains one old and one new strand. To another extent, it is also stochastic, namely because c) if the two strands of a chromosome in the mother cell differ in label concentration, it makes a difference for the daughter cell whether it receives the one or the other old strand, and d) due to heterogeneities in label uptake, strands that have been synthesized in distinct division events may differ in their label concentration. The mathematically rigorous description of this overall process turns out to be nontrivial, as will be seen in the following.

Let therefore the label concentration of a cell be denoted by the random variable X. Given the cell has L labeled strands, and $X_{\text{eff},s}$ is the effectively uptaken label in the s-th strand, then the label concentration is determined by

$$X = \sum_{s=1}^{L} X_{\text{eff},s}. \tag{6.5}$$

In order to describe the label distribution of cells, our aim is to derive the label densities

$$X \sim p(x|\mathbf{i}, j, t), \tag{6.6}$$

for which, according to the process of label uptake and inheritance as described above, we have to consider the number of labeled strands, as well as the effective label uptake into each labeled strand. Later, it will be seen that the label density depends only on the division numbers, i_1 and i_2, but not on the cell type j nor on the time t. This will allow for a decomposition approach to solve the model, as will also be proven later.

6.2.2 Modeling the DNA label segregation

According to (6.5), to obtain the label concentration X we need the number of labeled strands L, which is also a random variable. In this subsection, we turn as a first step towards modeling the number of labeled strands. As the cell type has no influence on the process of label uptake and inheritance, see (P1)–(P5) the number of labeled strands is independent of the cell type j. Because the number of labeled strands is not affected by any other processes than label uptake and inheritance, it is also independent of time t. The aim of this subsection is to mathematically describe the distribution of the number of labeled strands,

$$L \sim p_{\text{L}}(l|i_1, i_2), \tag{6.7}$$

determined by the number of divisions i_1 undergone during uplabeling, and i_2 during delabeling. The distribution $p_{\text{L}}(l|i_1, i_2)$ will be a probability mass function since the number of labeled strands L can only take discrete values $L = l \in \{0, \ldots, 2n_{\text{chrom}}\}$. It may be argued, in addition to the number of labeled strands, that chromosomes differ in size, which may contribute to the variability in label concentration. However, we discuss but do not explicitly incorporate size differences in this model, for reasons that will become clear in the discussion.

To the best of the author's knowledge, this is the first mathematically rigorous treatment of DNA label segregation from mother to daughter cells. The representation derived in the following may also cover other DNA labeling techniques, as they will usually depend on the segregation of chromosomes in a similar way. To derive a description of the number of labeled strands in the cell, let us consider the chromosomes and their state in terms of labeled strands. Each chromosome can be in one of the following three states:

- If none of the two DNA strands is labeled, then we call it an *unlabeled* chromosome.

- If one of the two DNA strands is labeled, then we call it a *one-fold labeled* chromosome.

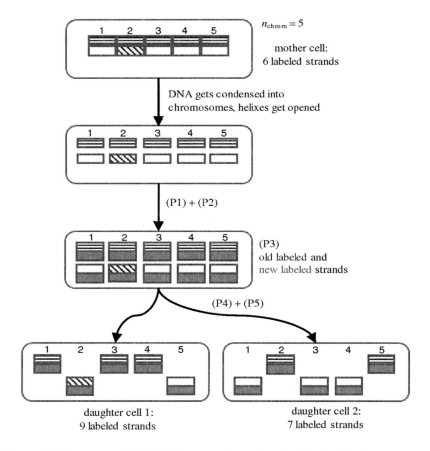

Figure 6.1: The process of DNA label segregation, as described in detail in the text, for an illustrative example of a cell with $n_{chrom} = 5$ chromosomes which divides during uplabeling. Each chromosome consists of a pair of DNA helix segments, which are denoted by rectangles. The DNA of the mother cell contains six labeled strands of which five have been synthesized in the same division event (denoted by horizontal lines), and one that has been synthesized in a distinct division event (diagonal lines). Before cell division, the DNA gets condensed into chromosomes and the DNA helixes get opened, thus the individual $2n_{chrom} = 10$ strands are now separated. The process of DNA label segregation upon cell division is described by (P1)–(P5) in the text. Finally, two daughter cells emerge out of the cell division process: The first daughter cell has nine labeled strands, which stem from in total three distinct division events (blue, red horizontal, red diagonal), while the second daughter cell has seven labeled strands that stem from two distinct division events (blue, red horizontal).

- If two of the two DNA strands are labeled, then we call it a *two-fold labeled* chromosome.

Thus we can describe each cell's labeling state by (L_0, L_1, L_2), where

- L_0 is the number of unlabeled chromosomes,

- L_1 is the number of one-fold labeled chromosomes, and

- L_2 is the number of two-fold labeled chromosomes.

In the illustrative example of Figure 6.1, the mother cell's labeling state is $(L_0, L_1, L_2) = (0, 4, 1)$. The total number of labeled strands in a cell is then given by

$$L = 0 \cdot L_0 + 1 \cdot L_1 + 2 \cdot L_2, \quad 0 \leq L \leq 2n_{\text{chrom}}, \tag{6.8}$$

whereas it always holds that

$$L_0 + L_1 + L_2 = n_{\text{chrom}}. \tag{6.9}$$

To describe the labeling state of cells in dependence of undergone divisions, we aim for a mathematical description of how the labeling state of the mother cell determines the labeling state of its daughter cells. This will provide a recursive definition of the labeling state, which, together with the base case of undivided cells, can be exploited to derive the desired distributions $p_{\text{L}}(l|i_1, i_2)$. Let therefore the mother cell's labeling state be denoted by $(L_0^{\text{mo}}, L_1^{\text{mo}}, L_2^{\text{mo}})$, and the labeling state of the two daughter cells by $(L_0^{\text{d1}}, L_1^{\text{d1}}, L_2^{\text{d1}})$ and $(L_0^{\text{d2}}, L_1^{\text{d2}}, L_2^{\text{d2}})$, respectively.

Serving as an intermediate step, we will formulate the joint probability distributions of cells having $L_0 = l_0$ unlabeled chromosomes, and $L_1 = l_1$ one-fold labeled chromosomes,

$$(L_0, L_1) \sim p_{\text{L01}}(l_0, l_1|i_1, i_2), \tag{6.10}$$

in dependence of undergone divisions. Since $L_2 = l_2$ is determined by (6.9) as $l_2 = n_{\text{chrom}} - l_0 - l_1$, let

$$\mathcal{I}_{\text{L01}} := \{l_0, l_1 | 0 \leq l_0 + l_1 \leq n_{\text{chrom}}\}, \tag{6.11}$$

denote the index set of all possible combinations of $L_0 = l_0$ and $L_1 = l_1$.

As will be pointed out in the following, unlabeled and two-fold labeled chromosomes in a mother cell yield chromosomes with ubiquitously determined labeling state in both daughter cells: Strands synthesized during uplabeling are always labeled, whereas strands synthesized during delabeling are always unlabeled. Therefore, all unlabeled chromosomes of the mother cell will result in one-fold labeled chromosomes for both daughter cells during uplabeling, and in unlabeled chromosomes for both daughter cells during delabeling (left column in Figure 6.2). Similarly, all two-fold labeled chromosomes of the mother cell will result again in two-fold labeled chromosomes for both daughter cells during uplabeling, and in one-fold labeled chromosomes during delabeling (right column in Figure 6.2).

The consequences of one-fold labeled chromosomes of the mother cell, in contrast, are less trivial: The labeled strand of such a chromosome may either go to daughter cell 1, or equally likely to daughter cell 2. If it goes to daughter cell 1, then daughter cell 1 will have one

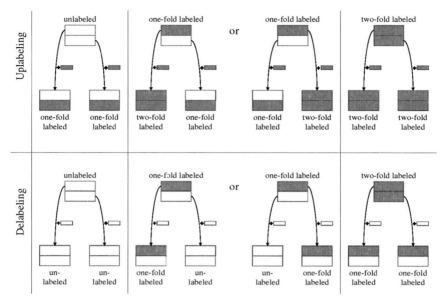

Figure 6.2: Chromosome segregation during uplabeling and delabeling. Uplabeling (upper row): An unlabeled chromosome results in one-fold labeled chromosomes in both daughter cells. A one-fold labeled chromosome results in a two-fold labeled chromosome in one daughter cell, and a one-fold labeled chromosome in the other daughter cell. A two-fold labeled chromosome results in two-fold labeled chromosomes in both daughter cells. Delabeling (lower row): An unlabeled chromosome results in unlabeled chromosomes in both daughter cells. A one-fold labeled chromosome results in a one-fold labeled chromosome in one daughter cell, and an unlabeled chromosome in the other daughter cell. A two-fold labeled chromosome results in one-fold labeled chromosomes in both daughter cells.

more labeled strand than daughter cell 2, and vice versa (middle column in Figure 6.2). This stochastic nature of the segregation of one-fold labeled chromosomes needs to be accounted for, since we desire to derive a mathematically rigorous model of DNA label segregation.

For this purpose, let the number of one-fold labeled chromosomes in the mother cell, of which the old labeled strand goes to daughter cell 1, be denoted by $C \in \{0, \ldots, n_{\text{chrom}}\}$. In the exemplary DNA label segregation process of Figure 6.1, one would have $C = 3$, and consequently $L_1^{\text{mo}} - C = 4 - 3 = 1$. The choice which one of the two strands of a chromosome goes to which daughter cell is purely random, and there are L_1^{mo} one-fold labeled chromosomes in the mother cell. Thus, the random variable C is sampled from a binomial distribution with L_1^{mo} trials and success probability 0.5,

$$C \sim \text{bino}(c | L_1^{\text{mo}}, 0.5). \tag{6.12}$$

Undivided during uplabeling   Cells that have not divided at all, $i_1 = i_2 = 0$, will have no labeled strands, thus have the labeling state $(L_0, L_1, L_2) = (n_{\text{chrom}}, 0, 0)$. Upon divisions during delabeling $i_2 \geq 1$, newly synthesized strands are always unlabeled, so this state will not change. With this, we know that if $i_1 = 0$, the joint probability distributions (6.10) are $\forall i_2 \geq 0$:

$$p_{\text{L01}}(l_0, l_1 | 0, 0) = \begin{cases} 1 & , l_0 = n_{\text{chrom}} \wedge l_1 = 0 \\ 0 & , \text{otherwise.} \end{cases} , \qquad (6.13)$$

and the distribution for the number of labeled strands (6.7) is $\forall i_2 \geq 0$:

$$p_{\text{L}}(l | 0, i_2) = \begin{cases} 1 & , l = 0 \\ 0 & , \text{otherwise.} \end{cases} . \qquad (6.14)$$

These cases $i_1 = 0, i_2 \geq 0$ will serve as the base cases for the recursive derivation in the following.

Divisions during uplabeling   Let us now consider cells that undergo divisions during uplabeling $(0, T_1]$, that is $i_1 \geq 1$ and $i_2 = 0$. The *number of unlabeled chromosomes* will in both daughter cells be zero since they conducted a division during the uplabeling phase:

$$L_0^{\text{d1}} = L_0^{\text{d2}} = 0. \qquad (6.15)$$

The *number of one-fold labeled chromosomes* is determined by the mother cell's L_0^{mo} unlabeled chromosomes, plus C of its one-fold labeled chromosomes, as introduced (6.12). According to the definition of C and because uplabeling adds a labeled strand, the one-fold labeled chromosomes in the mother cell yield $(L_1^{\text{mo}} - C)$ one-fold labeled chromosomes in daughter cell 1, and C one-fold labeled chromosomes in daughter cell 2. With this, the number of one-fold labeled chromosomes in the two daughter cells are given by

$$L_1^{\text{d1}} = L_0^{\text{mo}} + (L_1^{\text{mo}} - C), \qquad L_1^{\text{d2}} = L_0^{\text{mo}} + C. \qquad (6.16)$$

Note that during uplabeling, before the first division the number of unlabeled chromosomes will be $L_0^{\text{mo}} = n_{\text{chrom}}$ whereas there are no one-fold nor two-fold labeled chromosomes, $L_1^{\text{mo}} = L_2^{\text{mo}} = 0$. In contrast, after the first division there will be no more unlabeled chromosomes, thus always $L_0^{\text{mo}} = 0$. Using this, one may rewrite (6.16) for the first division event as $L_1^{\text{d1}} = n_{\text{chrom}}, L_1^{\text{d2}} = n_{\text{chrom}}$, and for all following division events, as $L_1^{\text{d1}} = L_1^{\text{mo}} - C, L_1^{\text{d2}} = C$.

Similarly, the *number of 2-fold labeled chromosomes* is determined by the mother cell's L_2^{mo} two-fold labeled chromosomes, plus C of its one-fold labeled chromosomes. According to the definition of C and because uplabeling adds a labeled strand, the one-fold labeled chromosomes in the mother cell yield C two-fold labeled chromosomes in daughter cell 1, and $(L_1^{\text{mo}} - C)$ two-fold labeled chromosomes in daughter cell 2. With this, the number of two-fold labeled chromosomes in the daughter cells are given by

$$L_2^{\text{d1}} = L_2^{\text{mo}} + C, \qquad L_2^{\text{d2}} = L_2^{\text{mo}} + (L_1^{\text{mo}} - C). \qquad (6.17)$$

This may again be specified, for the first division event: $L_2^{\text{d1}} = L_2^{\text{d2}} = 0$, for the second division event: $L_2^{\text{d1}} = C, L_2^{\text{d2}} = n_{\text{chrom}} - C$, whereas for all following division events as in (6.17).

With this we can now write the joint probability distributions (6.10) for a daughter cell having $L_0^d = l_0^d$ unlabeled chromosomes and $L_1^d = l_1^d$ one-fold labeled chromosomes, given i_1 divisions during uplabeling. We seek to express these probability distributions in dependence of the mother cell's respective probability distribution, and the binomial distribution for the variable C as in (6.12). Therefore we use the above derived expressions for $L_0^{d1}, L_0^{d2}, L_1^{d1}, L_1^{d2}$ to rewrite the expression for the value of C, whereas both possibilities of being daughter cell 1 or 2 are accounted, each with the probability 0.5. This yields the recursive definition $\forall i_1 \geq 1$:

$$
p_{\text{L01}}(l_0^d, l_1^d | i_1, 0) = \begin{cases} 0 & , l_0^d > 0 \\ 0.5 \displaystyle\sum_{\mathcal{I}_{\text{L01}}} p_{\text{L01}}(l_0^{mo}, l_1^{mo} | i_1 - 1, 0)\text{bino}(-l_1^d + l_0^{mo} + l_1^{mo} | l_1^{mo}, 0.5) \\ \quad + 0.5 \displaystyle\sum_{\mathcal{I}_{\text{L01}}} p_{\text{L01}}(l_0^{mo}, l_1^{mo} | i_1 - 1, 0)\text{bino}(l_1^d - l_0^{mo} | l_1^{mo}, 0.5) & , l_0^d = 0 \end{cases}
$$

$$
= \begin{cases} 0 & , l_0^d > 0 \\ \displaystyle\sum_{\mathcal{I}_{\text{L01}}} p_{\text{L01}}(l_0^{mo}, l_1^{mo} | i_1 - 1, 0)\text{bino}(l_1^d - l_0^{mo} | l_1^{mo}, 0.5) & , l_0^d = 0 \end{cases} ,
$$

$$(6.18)$$

with the index set $\mathcal{I}_{\text{L01}}$ as in (6.11), and the base case for $i_1 = 0$ as given in (6.13).

Divisions during delabeling Now, let us pursue the same approach considering cells that undergo divisions during delabeling $(T_1, T_2]$, that is $i_2 \geq 1$ for any $i_1 \geq 0$. The *number of two-fold labeled chromosomes* will in both daughter cells be zero, since they conducted at least one division during the delabeling phase, thus

$$L_2^{d1} = L_2^{d2} = 0. \qquad (6.19)$$

The *number of one-fold labeled chromosomes* depends on the mother cell's L_2^{mo} two-fold labeled chromosomes, plus C of its one-fold labeled chromosomes, as introduced before, (6.12). According to the definition of C, and because delabeling adds an unlabeled strand to the inherited strand, the one-fold labeled chromosomes in the mother cell yield C one-fold labeled chromosomes in daughter cell 1, and $(L_1^{mo} - C)$ one-fold labeled chromosomes in daughter cell 2. With this, the number of one-fold labeled chromosomes in the daughter cells can be expressed as

$$L_1^{d1} = L_2^{mo} + C, \quad L_1^{d2} = L_2^{mo} + (L_1^{mo} - C). \qquad (6.20)$$

In a similar way, the *number of unlabeled chromosomes* depends on the mother cell's L_0^{mo} unlabeled chromosomes, plus C of its one-fold labeled chromosomes. Because delabeling adds an unlabeled strand to the inherited strand, the one-fold labeled chromosomes in the mother cell yield $(L_1^{mo} - C)$ unlabeled chromosomes in daughter cell 1, and C unlabeled chromosomes in daughter cell 2. With this, the number of unlabeled chromosomes in the daughter cells are given by

$$L_0^{d1} = L_0^{mo} + (L_1^{mo} - C), \quad L_0^{d2} = L_0^{mo} + C. \qquad (6.21)$$

With this we are ready to write the joint probability distributions (6.10) for a daughter cell having $L_0^d = l_0^d$ unlabeled chromosomes and $L_1^d = l_1^d$ one-fold labeled chromosomes, thus

$L_2^d = n_{\text{chrom}} - l_0^d - l_1^d$ two-fold labeled chromosomes. This yields again a recursive definition $\forall i_2 \geq 1, i_1 \geq 0$:

$$p_{\text{L01}}(l_0^d, l_1^d | i_1, i_2) = \begin{cases} 0 & , l_2^d > 0 \\ 0.5 \sum\limits_{\mathcal{I}_{\text{L01}}} p_{\text{L01}}(l_0^{mo}, l_1^{mo} | i_1, i_2 - 1) \text{bino}(-l_0^d + l_0^{mo} + l_1^{mo} | l_1^{mo}, 0.5) \\ \quad + 0.5 \sum\limits_{\mathcal{I}_{\text{L01}}} p_{\text{L01}}(l_0^{mo}, l_1^{mo} | i_1, i_2 - 1) \text{bino}(l_0^d - l_0^{mo} | l_1^{mo}, 0.5) & , l_2^d = 0 \end{cases}$$

$$= \begin{cases} 0 & , l_2^d > 0 \\ \sum\limits_{\mathcal{I}_{\text{L01}}} p_{\text{L01}}(l_0^{mo}, l_1^{mo} | i_1, i_2 - 1) \text{bino}(l_0^d - l_0^{mo} | l_1^{mo}, 0.5) & , l_2^d = 0 \end{cases} ,$$

$$(6.22)$$

with the index set $\mathcal{I}_{\text{L01}}$ as in (6.11), and the base cases given by (6.13).

Using the obtained results for the joint probability distributions (6.13), (6.18), and (6.22), we can go back to deriving the originally desired probability mass function (6.7). For the total number of labeled strands l as in (6.8), we get the distributions

$$L \sim p_{\text{L}}(l | i_1, i_2) = \sum_{\substack{\{l_0, l_1 | 0 \cdot l_0 + 1 \cdot l_1 \\ + 2 \cdot (n_{\text{chrom}} - l_0 - l_1) = l\}}} p_{\text{L01}}(l_0, l_1 | i_1, i_2). \tag{6.23}$$

Exemplary probability mass functions $p_{\text{L}}(l | i_1, i_2)$, in dependence of divisions i_1 during uplabeling and i_2 during delabeling, for human cells with $n_{\text{chrom}} = 46$, are shown in Figure 6.3. It is seen that after one division in uplabeling, $i_1 = 1$, the distribution is a peak at $l = n_{\text{chrom}} = 46$, since exactly half of the strands will be labeled. For higher division numbers $i_1 > 1$ however, the probability distribution is spread over a range of possible values l for the number of labeled strands.

The individual probability distributions over l_0, l_1, and l_2 may be achieved from this via marginalizations, that is summing over all possible corresponding states:

$$p_{\text{L0}}(l_0 | i_1, i_2) = \sum_{l_1} p_{\text{L01}}(l_0, l_1 | i_1, i_2),$$

$$p_{\text{L1}}(l_1 | i_1, i_2) = \sum_{l_0} p_{\text{L01}}(l_0, l_1 | i_1, i_2), \tag{6.24}$$

$$p_{\text{L2}}(l_2 | i_1, i_2) = \sum_{\substack{\{l_0, l_1 | \\ n_{\text{chrom}} - l_0 - l_1 = l_2\}}} p_{\text{L01}}(l_0, l_1 | i_1, i_2).$$

Although the probability distribution for the number of labeled strands that was derived in this subsection represents only an intermediate result towards modeling the label concentration of cells, we can already draw some conclusions, regarding the DNA label segregation. In division numbers $i_1 \geq 2$, the process of chromosome segregation alone induces already a distribution of label content (opposed to the results when assuming a fixed label content, as for example by Ganusov and De Boer (2013)). For the first division number $i_1 = 1$, however, the here presented model predicts a sharp peak at $l = n_{\text{chrom}}$, meaning that after the first division,

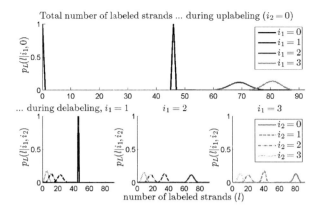

Figure 6.3: Probability distributions $p_L(l|i_1, i_2)$ for the total number of labeled strands l during uplabeling (upper row) and during delabeling (lower row), in dependence of the division numbers i_1, i_2, exemplary for human cells with $n_{chrom} = 46$ chromosomes. The resulting distributions (probability masses) $p_L(l|i_1, i_2)$ are shown for selected values of division numbers i_1 during uplabeling, and i_2 during delabeling.

exactly half of the strands will be labeled with probability one. This is intuitive but in contrast to data that show BrdU intensities of cells that have undergone no or one division during uplabeling (Takizawa et al., 2011). This result would not change even if different chromosome sizes were considered in the model, because each chromosome then has exactly one unlabeled and one labeled strand. In conclusion, neither the process of chromosome segregation, nor differences in chromosome size, can be the only source of the reported heterogeneity in measured label intensities.

To further develop the model for BrdU-labeled cells, we will include a heterogeneous labeling efficacy in Subsection 6.2.3, as well as the autofluorescence in Subsection 6.2.5. This will then also serve to investigate the potential of these two contributions for being a source of heterogeneity in measured label intensities.

6.2.3 Modeling the label concentration

After we have determined the number of labeled chromosome strands L, we will as the next step combine this with the labeling efficacy X_{eff} in order to derive the label concentration X, as given in (6.5). Such a labeling efficacy captures uncertainties during the process of BrdU incorporation, as they may arise for example from heterogeneities in environmental labeling conditions within an organism.

Let us therefore assume that the effective label uptake, X_{eff}, is sampled from a lognormal distribution,

$$X_{eff} \sim \log\mathcal{N}(x_{eff}|\mu_{eff}, \sigma_{eff}^2). \tag{6.25}$$

As outlined in Subsection 6.2.1, it is assumed that upon cell division, all newly labeled chromosome strands receive the same label concentration. This is reflected by the following property: For each division i_1 during uplabeling, there is one realization of the random variable, $X_{\text{eff}}(i_1)$, which represents the effective label uptake into one newly synthesized strand for all strands j synthesized in this division event, $X_{\text{eff},j} = X_{\text{eff}}(i_1)$. For two different division events $i_1^{(1)}$, $i_1^{(2)}$ during uplabeling, the according two random variables $X_{\text{eff}}(i_1^{(1)})$ and $X_{\text{eff}}(i_1^{(1)})$ are independent identically distributed.

The number of labeled strands, L, was seen in the previous subsection to depend on the number of undergone divisions according to the probability mass function (6.23), which can be achieved recursively. This probability mass function is defined over discrete values $l \in \{0, \ldots, 2n_{\text{chrom}}\}$, with n_{chrom} the number of chromosomes of the cell. The properties that $p_{\text{L}}(l|i_1, i_2)$ is a probability mass function and that it can be computed analytically by (6.23) will be used in the following to combine the number of labeled strands with the effectively uptaken label, which together yields the label concentration.

Now, we aim to derive an expression for the label density $p(x|i_1, i_2)$, (6.6). We will treat three subsequent division cases separately: Undivided during uplabeling, divisions during uplabeling, and divisions during delabeling.

Undivided during uplabeling Cells that have not divided during uplabeling will have no labeled strands at all, $X(0, i_2) = 0$, regardless of divisions during delabeling. Consequently, we have for $i_1 = 0, \forall i_2 \geq 0$:

$$p(x|0, i_2) = \delta(x). \tag{6.26}$$

Divisions during uplabeling During the uplabeling phase $t \in (0, T_1]$, each i_1-th division event adds n_{chrom} newly labeled strands, and $(L - n_{\text{chrom}})$ old labeled strands are inherited from the mother cell. So the first part contributing to the label concentration of the new cell will be made up by n_{chrom}-times the same random variable,

$$n_{\text{chrom}} X_{\text{eff}}(i_1) \sim \frac{1}{n_{\text{chrom}}} \log \mathcal{N} \left(\frac{1}{n_{\text{chrom}}} x_{\text{eff}} \middle| \mu_{\text{eff}}, \sigma_{\text{eff}}^2 \right)$$
$$= \log \mathcal{N} \left(x_{\text{eff}} | \mu_{\text{eff}} + \log(n_{\text{chrom}}), \sigma_{\text{eff}}^2 \right). \tag{6.27}$$

The second part contributing to the label concentration concerns the inherited $(L - n_{\text{chrom}})$ strands. These may stem in general partly from same or/and distinct division events. Summing up the effective label uptake of all individual old strands, $\sum_{j=1}^{L-n_{\text{chrom}}} X_{\text{eff}}(i_1^{(j)})$, would require to account for all possible combinations of these strands originating from various division events $i_1^{(j)}$, giving up to $(L - n_{\text{chrom}})^{i_1}$ possible combinations. Obviously, this would result in an exploding complexity of the model. Fortunately, we can replace the exact calculation by an approximation based on the following simplifying assumption:

(A1) Each of the old labeled strands is assumed to stem from a distinct division event, thus their effective label uptake is given by $(L - n_{\text{chrom}})$ independent identically distributed random variables.

We have evaluated the impact of this assumption by comparing stochastic simulations of the full process, as given by (P1)–(P5), against stochastic simulations of the process assuming (A1), each with 10000 dividing cells. The results are shown in Figure 6.4, from which it is seen that the approximation is in good agreement with the exact process. Together with the fact that the assumption (A1) clearly reduces the complexity, by reducing the possible combinations from maximally $(L - n_{\text{chrom}})^{i_1}$ to $(L - n_{\text{chrom}})$, this justifies to proceed assuming (A1).

Because the $(L - n_{\text{chrom}})$ inherited strands are now regarded as independent random variables, the distribution of the sum $\sum_{j=1}^{L - n_{\text{chrom}}} X_{\text{eff},j}$ is a convolution of $(L - n_{\text{chrom}})$ lognormal distributions,

$$\sum_{j=1}^{L - n_{\text{chrom}}} X_{\text{eff},j} \sim \text{conv}_{j=1}^{L - n_{\text{chrom}}} \log \mathcal{N}\left(x_{\text{eff}} | \mu_{\text{eff}}, \sigma_{\text{eff}}^2\right), \tag{6.28}$$

with conv denoting the recursive convolution of its arguments,

$$\text{conv}_{\iota=1,2}\left(p_{\iota}(x)\right) := \int_{\mathbb{R}_+} p_1(\xi) p_2(x - \xi) d\xi. \tag{6.29}$$

With summing up the two parts (6.27) and (6.28), the label concentration during uplabeling is distributed according to

$$X(i_1, 0) = n_{\text{chrom}} X_{\text{eff}}(i_1, 0) + \sum_{j=1}^{L - n_{\text{chrom}}} X_{\text{eff},j}$$

$$\sim \int_{\mathbb{R}_+} \log \mathcal{N}(x - \chi | \mu_{\text{eff}} + \log(n_{\text{chrom}}), \sigma_{\text{eff}}^2) \, \text{conv}_{j=1}^{L - n_{\text{chrom}}} \log \mathcal{N}\left(\chi | \mu_{\text{eff}}, \sigma_{\text{eff}}^2\right) d\chi, \tag{6.30}$$

with $L \sim p_{\text{L}}(l | i_1, 0)$. In order to get rid of the dependency on the random variable L, we aim to express the label concentration in dependence of the division number. This can be achieved as follows: We know that the number of undergone divisions i_1 determines the probability of having a certain number $L = l$ of labeled strands, as given in (6.23). For each possible value of the random variable L (number of labeled chromosome strands), $L = l \in \{0, \dots, 2n_{\text{chrom}}\}$, we write the probability of having this specific value l, determining the weights $p_{\text{L}}(l | i_1, i_2)$, and multiply it with the probability density of label concentration, given this l, (6.30), and sum these up over all possible $l \in \{0, \dots, 2n_{\text{chrom}}\}$. Then $\forall i_1 \geq 1, i_2 = 0$:

$$p(x | i_1, 0) \overset{(6.30)}{=} \int_{\mathbb{R}_+} \log \mathcal{N}(x - \chi | \mu_{\text{eff}} + \log(n_{\text{chrom}}), \sigma_{\text{eff}}^2)$$

$$\cdot \sum_{l=0}^{2n_{\text{chrom}}} p_{\text{L}}(l | i_1, 0) \text{conv}_{j=1}^{l - n_{\text{chrom}}} \log \mathcal{N}\left(\chi | \mu_{\text{eff}}, \sigma_{\text{eff}}^2\right) d\chi. \tag{6.31}$$

The first part, (6.27), represents the label uptake in the recent division event, and thus this gives the label uptake during uplabeling as in (6.3),

$$\forall t \in (0, T_1]: \quad p_{\text{lab}}(x - \chi | t) = \log \mathcal{N}(x - \chi | \mu_{\text{eff}} + \log(n_{\text{chrom}}), \sigma_{\text{eff}}^2). \tag{6.32}$$

The second part, (6.28), in turn represents the label inheritance, $\forall i_1 \geq 1, i_2 = 0$:

$$p_{\text{inhe}}(\chi | (i_1 - 1, 0), \mu, t) = \sum_{l=0}^{n_{\text{chrom}}} p_{\text{L}}(l | i_1, i_2) \text{conv}_{j=1}^{l - n_{\text{chrom}}} \log \mathcal{N}\left(\chi | \mu_{\text{eff}}, \sigma_{\text{eff}}^2\right). \tag{6.33}$$

With this we have achieved the label uptake and label inheritance distributions as they appear in the DsCL model equations, (4.1) for the uplabeling phase.

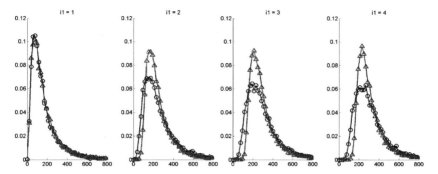

Figure 6.4: Histograms of label concentrations of 10000 dividing cells, stochastically simulated with using the full process ($\circ$), using the assumption (A1) ($\triangle$), depicted for the division numbers $i_1 = 1, \ldots, 4$.

Divisions during delabeling At the end of the uplabeling phase $t = T_1$, cells have $L = L(i_1, 0)$ labeled strands. Upon divisions during delabeling, these $L(i_1, 0)$ labeled strands get subsequently segregated to daughter cells, such that after i_2 divisions the number of labeled strands will be $L(i_1, i_2) \leq L(i_1, 0)$. Of all these labeled strands, a certain number of strands U will originate from the last division during uplabeling. Upon each division during delabeling, there is an equal chance for each chromosome to inherit one or the other strand, thus the number of strands that stem from the last uplabeling division is

$$U \sim p_{\mathrm{U}}(u|i_2) = \mathrm{bino}(u|n_{\mathrm{chrom}}, 0.5^{i_2}). \tag{6.34}$$

In addition, we carry on the assumption (A1) that all strands labeled before the last uplabeling division event have independent values, and only the n_{chrom} strands labeled in the last uplabeling division event i_1 have the same value of effective label uptake. With this, we can express the first part contributing to the label concentration by

$$\begin{aligned} U X_{\mathrm{eff}}(i_1) &\sim \frac{1}{U} \log \mathcal{N} \left(\frac{1}{U} x_{\mathrm{eff}} \middle| \mu_{\mathrm{eff}}, \sigma_{\mathrm{eff}}^2 \right) \\ &= \log \mathcal{N} \left(x_{\mathrm{eff}} \middle| \mu_{\mathrm{eff}} + \log(U), \sigma_{\mathrm{eff}}^2 \right). \end{aligned} \tag{6.35}$$

The second part of the label concentration in turn is

$$\sum_{j=1}^{L-U} X_{\mathrm{eff},j} \sim \mathrm{conv}_{j=1}^{L-U} \log \mathcal{N} \left(x_{\mathrm{eff}} \middle| \mu_{\mathrm{eff}}, \sigma_{\mathrm{eff}}^2 \right). \tag{6.36}$$

With summing up the two parts (6.35) and (6.36), we obtain

$$\begin{aligned} X(i_1, i_2) &= U X_{\mathrm{eff}}(i_1, i_2) + \sum_{j=1}^{L-U} X_{\mathrm{eff},j} \\ &\sim \int_{\mathbb{R}_+} \log \mathcal{N}(x - \chi | \mu_{\mathrm{eff}} + \log(U), \sigma_{\mathrm{eff}}^2) \, \mathrm{conv}_{j=1}^{L-U} \log \mathcal{N} \left(\chi | \mu_{\mathrm{eff}}, \sigma_{\mathrm{eff}}^2 \right) d\chi, \end{aligned} \tag{6.37}$$

with $L \sim p_{\mathrm{L}}(l|i_1, i_2)$ and $U \sim p_{\mathrm{U}}(u|i_2)$. We aim to dispose again of the dependency on the random variables L and U. Similarly to the approach conducted for the uplabeling phase, we can sum over all possible values for these random variables and achieve $\forall i_1 > 0, i_2 \geq 0$:

$$p(x|i_1, i_2) \overset{(6.38)}{=} \sum_{l=0}^{2n_{\mathrm{chrom}}} \sum_{u=0}^{2n_{\mathrm{chrom}}} p_{\mathrm{L}}(l|i_1, i_2)\, p_{\mathrm{U}}(u|i_2) \int_{\mathbb{R}_+} \log \mathcal{N}(x - \chi|\mu_{\mathrm{eff}} + \log(u), \sigma_{\mathrm{eff}}^2)$$
$$\cdot \operatorname{conv}_{j=1}^{l-u} \log \mathcal{N} \left(\chi|\mu_{\mathrm{eff}}, \sigma_{\mathrm{eff}}^2 \right) d\chi.$$
$$(6.38)$$

Since during delabeling there is no label uptake, (6.4), one concludes that (6.37) gives the label inheritance distribution, $\forall i_1 \geq 1, i_2 \geq 1$:

$$p_{\mathrm{inhe}}(x|(i_1, i_2 - 1), \mu, t) = p(x|i_1, i_2). \qquad (6.39)$$

With this we have also obtained the label uptake and label inheritance distributions as they appear in the DsCL model equations, (4.1) for the delabeling phase.

With (6.26), (6.31), and (6.38) we have now expressed the label densities (6.6), whereas the assumption (A1) was exploited for an approximation of clearly reduced complexity. As an important side result, it was seen in this subsection that the label densities (6.6) do not depend on cell types j, nor on time t, but solely on the division numbers $\mathbf{i} = (i_1, i_2)$. These facts can be exploited for a decomposition approach in order to derive $n(x, \mathbf{i}, j|t)$ and thus the overall label distribution $m(x|t)$. We will present this solution approach in the next subsection. An approach to efficiently calculate the distributions, while considering also the autofluorescence in addition, will then be developed in the subsection thereafter.

6.2.4 Model solution

In the previous section, it was seen that for BrdU labeling, the label densities (6.6) can be written as $\forall (\mathbf{i}, t) \in \mathbb{N}_0^2 \times (T_{k-1}, T_k]$:

$$p(x|\mathbf{i}) = \int_0^\infty p_{\mathrm{lab}}(x - \chi|t) p_{\mathrm{inhe}}(\chi|\mathbf{i} - \mathbf{e}_k^T, \mu, t) d\chi$$
$$= \int_0^\infty p_{\mathrm{lab}}^{(k)}(x - \chi) p_{\mathrm{inhe}}(\chi|\mathbf{i} - \mathbf{e}_k^T) d\chi, \qquad (6.40)$$

with $\forall t : p_{\mathrm{lab}}(x|t) = p_{\mathrm{lab}}^{(k)}(x)$ the label uptake as in (6.32) and (6.4) which is invariant during each time interval $(T_{k-1}, T_k]$, and $p_{\mathrm{inhe}}(x|\mathbf{i} - \mathbf{e}_k^T)$ the label inheritance as in (6.33) and (6.39) which solely depends on the number of undergone divisions.

We can insert this together with the specifications in Subsection 6.2.1 into the DsCL model equations (4.1). Then, we get the following model for the dynamics of $n(x, \mathbf{i}, j|t)$, given by a system of PDEs, $\forall \mathbf{i} = (i_1, i_2) \in \mathbb{N}_0^2, \forall j \in \{1, \ldots, J\}, \forall k \in \{1, 2\}, \forall t \in (T_{k-1}, T_k]$:

$$
\begin{aligned}
\frac{\partial n(x, \mathbf{i}, j | t)}{\partial t} = & -\sum_{\mu=1}^{J} \delta_{\mathbf{i}}^{\mu j}(t) n(x, \mathbf{i}, j | t) + \sum_{\mu=1}^{J} \delta_{\mathbf{i}}^{j\mu}(t) n(x, \mathbf{i}, \mu | t) \\
& - \beta_{\mathbf{i}}^{j}(t) n(x, \mathbf{i}, j | t) - \alpha_{\mathbf{i}}^{j}(t) n(x, \mathbf{i}, j | t) \\
& + \begin{cases} 0 & , i_k = 0 \\ 2 \sum_{\mu=1}^{J} \alpha_{\mathbf{i}-\mathbf{e}_k^T}^{\mu}(t) w_{\mathbf{i}-\mathbf{e}_k^T}^{j\mu}(t) \\ \quad \cdot \int_{\mathbb{R}_+} p_{\text{lab}}^{(k)}(x-\chi) p_{\text{inhe}}(\chi | \mathbf{i} - \mathbf{e}_k^T) \int_{\mathbb{R}_+} n(\zeta, \mathbf{i} - \mathbf{e}_k^T, \mu, t) d\zeta \, d\chi & , i_k \geq 1 \end{cases}
\end{aligned}
$$

$$\tag{6.41}$$

with initial conditions

$$
n(x, (0,0), j | 0) = N_{\text{ini},j} \delta(x), \ \forall \mathbf{i} \neq (0,0) : n(x, \mathbf{i}, j | 0) = 0. \tag{6.42}
$$

Decomposability of the model From (6.40) it follows directly that (4.11) holds, which was proven in Theorem 4.2 to be a sufficient condition for the decomposability (4.12) of the model. We can therefore exploit the decomposition approach, which allows for an efficient solution scheme that circumvents the need to solve the system of coupled PDEs (6.41).

This is captured by the following corollary:

Corrolary 6.1 *The solution $n(x, \mathbf{i}, j | t)$ of the model (6.41) is given by*

$$
n(x, \mathbf{i}, j | t) = N(\mathbf{i}, j | t) p(x | \mathbf{i}), \tag{6.43}
$$

whereas $N(\mathbf{i}, j | t)$ is the solution of the system of ODEs $\forall \mathbf{i} = (i_1, i_2) \in \mathbb{N}_0^2, j \in \{1, \ldots, J\}, k \in \{1, 2\}, \forall t \in (T_{k-1}, T_k]$:

$$
\begin{aligned}
\frac{dN(\mathbf{i}, j | t)}{dt} = & -\sum_{\mu=1}^{J} \delta_{\mathbf{i}}^{\mu j}(t) N(\mathbf{i}, j | t) + \sum_{\mu=1}^{J} \delta_{\mathbf{i}}^{j\mu}(t) N(\mathbf{i}, \mu | t) \\
& - \beta_{\mathbf{i}}^{j}(t) N(\mathbf{i}, j | t) - \alpha_{\mathbf{i}}^{j}(t) N(\mathbf{i}, j | t) \\
& + \begin{cases} 0 & , i_k = 0 \\ 2 \sum_{\mu=1}^{J} \alpha_{\mathbf{i}-\mathbf{e}_k^T}^{\mu}(t) w_{\mathbf{i}-\mathbf{e}_k^T}^{j\mu}(t) N(\mathbf{i} - \mathbf{e}_k^T, \mu | t) & , i_k \geq 1 \end{cases}
\end{aligned}
$$

$$\tag{6.44}$$

with initial conditions

$$
\forall j : N((0,0), j | 0) = N_{\text{ini},j}, \forall \mathbf{i} \neq (0,0) : N(\mathbf{i}, j | 0) = 0, \tag{6.45}
$$

and $p(x | \mathbf{i})$ is defined by (6.40).

The quantities $N(\mathbf{i}, j | t)$ give the numbers of cells in each subpopulation (4.3), whereas the quantities $p(x | \mathbf{i})$ give the label densities in each subpopulation (4.8).

The decomposability of the model is a powerful result for the analysis and simulation of the model, as we can now pursue the solution $n(x, \mathbf{i}, j | t)$ of the model from the decomposed parts: The dynamics of the number of cells, $N(\mathbf{i}, j|t)$, can be obtained by solving the system of ODEs (6.44). Depending on the complexity of parameters, one may therefore exploit either an analytical solution, or a truncated ODE system, which can be solved numerically. A detailed discussion of these options was provided in Section 4.2.2. The probability densities $p(x|\mathbf{i})$ in turn can be calculated as presented in the previous subsection. Then, the overall solution can be reassembled as $n(x, \mathbf{i}, j|t) = N(\mathbf{i}, j|t)p(x|\mathbf{i})$.

6.2.5 Modeling the measured label distribution

In the measurement output of labeling experiments, one actually does not obtain the overall label distribution $m(x|t)$ directly, but the values of fluorescence induced by the actual label concentration and of additional autofluorescence. Let the autofluorescence be denoted by a random variable, Y_a, which is assumed to follow a lognormal distribution,

$$Y_a \sim \log \mathcal{N}(y_a|\mu_a, \sigma_a^2). \tag{6.46}$$

The label concentration X induces a fluorescence, which we assume here to equal the label concentration X. This can be easily justified since the labeling efficacy, and thus the label concentration, could already include the respective scaling, but more generally cX could also be substituted into the model. A more detailed discussion on modeling the dynamics of the induced fluorescence can be found in Hasenauer et al. (2012a). With this, the measured label intensity consists of the sum of label concentration and autofluorescence, for $\mathbf{i} = (i_1, i_2)$,

$$Y = X + Y_a \sim \hat{p}(y|\mathbf{i}). \tag{6.47}$$

Then, the actual measurement output is given by the measured label distribution

$$\hat{m}(y|t) = \sum_{i_1=0}^{\infty} \sum_{i_2=0}^{\infty} \sum_{j=1}^{J} \hat{n}(y, \mathbf{i}, j|t) = \sum_{i_1=0}^{\infty} \sum_{i_2=0}^{\infty} \sum_{j=1}^{J} \int_{\mathbb{R}_+} n(\zeta, \mathbf{i}, j|t) \log \mathcal{N}(y - \zeta|\mu_a, \sigma_a^2) d\zeta \tag{6.48}$$

Due to the decomposition property (6.43) of the model, the individual label distributions $\hat{n}(y, \mathbf{i}, j|t)$ can also be decomposed,

$$\hat{n}(y, \mathbf{i}, j|t) = \int_{\mathbb{R}_+} n(\zeta, \mathbf{i}, j|t) \log \mathcal{N}(y - \zeta|\mu_a, \sigma_a^2) d\zeta$$
$$= \int_{\mathbb{R}_+} N(\mathbf{i}, j|t)p(\zeta|\mathbf{i}) \log \mathcal{N}(y - \zeta|\mu_a, \sigma_a^2) d\zeta = N((i_1, i_2), j|t)\hat{p}(y|i_1, i_2), \tag{6.49}$$

with $\hat{p}(y|i_1, i_2)$ the probability density for the measured label intensity (6.47). We now aim to derive $\hat{p}(y|i_1, i_2)$, for which we will finally present an efficient yet highly accurate approximation.

Undivided during uplabeling If no divisions during uplabeling were conducted, the measured label density equals, as expected, the autofluorescence distribution, for $i_1 = 0, \forall i_2 \geq 0$:

$$\hat{p}(y|0, i_2) = \int_{\mathbb{R}_+} \delta(\zeta) \log \mathcal{N}(y - \zeta|\mu_a, \sigma_a^2) d\zeta \overset{(6.46)}{=} \log \mathcal{N}(y|\mu_a, \sigma_a^2). \tag{6.50}$$

Divisions during uplabeling After divisions during uplabeling, using (6.31) and (6.46), the measured label densities are $\forall i_1 \geq 1, i_2 = 0$:

$$
\begin{aligned}
\hat{p}(y|i_1, 0) &= \int_{\mathbb{R}_+} p(\zeta|i_1, i_2) \log \mathcal{N}(y - \zeta|\mu_\mathrm{a}, \sigma_\mathrm{a}^2) d\zeta \\
&= \int_{\mathbb{R}_+} \int_{\mathbb{R}_+} \log \mathcal{N}(\zeta - \chi|\mu_\mathrm{eff} + \log(n_\mathrm{chrom}), \sigma_\mathrm{eff}^2) \\
&\qquad\qquad \cdot \sum_{l=0}^{2n_\mathrm{chrom}} \left(p_\mathrm{L}(l|i_1, 0) \mathrm{conv}_{j=1}^{l-n_\mathrm{chrom}} \log \mathcal{N}\left(\chi|\mu_\mathrm{eff}, \sigma_\mathrm{eff}^2\right) \right) d\chi \, \log \mathcal{N}(y - \zeta|\mu_\mathrm{a}, \sigma_\mathrm{a}^2) d\zeta \\
&= \sum_{l=0}^{2n_\mathrm{chrom}} p_\mathrm{L}(l|i_1, 0) \int_{\mathbb{R}_+} \int_{\mathbb{R}_+} \log \mathcal{N}(\zeta - \chi|\mu_\mathrm{eff} + \log(n_\mathrm{chrom}), \sigma_\mathrm{eff}^2) \\
&\qquad\qquad \cdot \mathrm{conv}_{j=1}^{l-n_\mathrm{chrom}} \log \mathcal{N}\left(\chi|\mu_\mathrm{eff}, \sigma_\mathrm{eff}^2\right) d\chi \, \log \mathcal{N}(y - \zeta|\mu_\mathrm{a}, \sigma_\mathrm{a}^2) d\zeta.
\end{aligned}
\tag{6.51}
$$

This expression contains $(2n_\mathrm{chrom}+1)$ summands each with $(l-n_\mathrm{chrom}+2)$ nested convolution integrals. The calculation of (6.51) would in total require the evaluation of $(n^2+5n+4)/2$ convolution integrals, which induces a tremendous computational burden and additional sources for numerical errors. Especially the large computational burden might be not feasible, for example when using the model for exhaustive simulations or parameter estimation. Fortunately, we can exploit the fact that all involved distributions are lognormal distributions, and that according to Fenton (1960) convolutions of lognormal distributions can be well approximated by a lognormal distribution via a moment-matching approach: The parameters of the approximating lognormal distribution will be chosen such that its mean and variance equal that of the actual exact convolution. We use this approach (Fenton, 1960) to approximate the individual summands in (6.51) as follows:

$$
\begin{aligned}
&\int_{\mathbb{R}_+} \int_{\mathbb{R}_+} \log \mathcal{N}(\zeta - \chi|\mu_\mathrm{eff} + \log(n_\mathrm{chrom}), \sigma_\mathrm{eff}^2) \\
&\qquad\qquad \cdot \mathrm{conv}_{j=1}^{l-n_\mathrm{chrom}} \log \mathcal{N}\left(\chi|\mu_\mathrm{eff}, \sigma_\mathrm{eff}^2\right) d\chi \, \log \mathcal{N}(y - \zeta|\mu_\mathrm{a}, \sigma_\mathrm{a}^2) d\zeta \\
&\approx \log \mathcal{N}\left(y|\tilde{\mu}^{(l)}, (\tilde{\sigma}^{(l)})^2\right),
\end{aligned}
\tag{6.52}
$$

where the parameters of the approximating lognormal distribution are

$$
\begin{aligned}
\tilde{\mu}^{(l)} &= \log\left(\mathrm{E}_\mathrm{a} + \mathrm{E}_\mathrm{C} + \mathrm{E}_\mathrm{L}^{(l)}\right) - \frac{1}{2} \log\left(\frac{\mathrm{Var}_\mathrm{a} + \mathrm{Var}_\mathrm{C} + \mathrm{Var}_\mathrm{L}^{(l)}}{(\mathrm{E}_\mathrm{a} + \mathrm{E}_\mathrm{C} + \mathrm{E}_\mathrm{L}^{(l)})^2} + 1\right) \\
\tilde{\sigma}^{(l)} &= \sqrt{\log\left(\frac{\mathrm{Var}_\mathrm{a} + \mathrm{Var}_\mathrm{C} + \mathrm{Var}_\mathrm{L}^{(l)}}{(\mathrm{E}_\mathrm{a} + \mathrm{E}_\mathrm{C} + \mathrm{E}_\mathrm{L}^{(l)})^2} + 1\right)}.
\end{aligned}
\tag{6.53}
$$

121

For this, the first two moments of the individual distributions are used to derive

$$
\begin{aligned}
\mathrm{E_a} &= e^{\mu_\mathrm{a}+\frac{\sigma_\mathrm{a}^2}{2}} \\
\mathrm{E_C} &= n_{\mathrm{chrom}}e^{\mu_{\mathrm{eff}}+\frac{\sigma_{\mathrm{eff}}^2}{2}} \\
\mathrm{E_L^{(l)}} &= (l - n_{\mathrm{chrom}})e^{\mu_{\mathrm{eff}}+\frac{\sigma_{\mathrm{eff}}^2}{2}} \\
\mathrm{Var_a} &= e^{2\mu_\mathrm{a}+\sigma_\mathrm{a}^2}\left(e^{\sigma_\mathrm{a}^2}-1\right) \\
\mathrm{Var_C} &= n_{\mathrm{chrom}}^2 e^{2\mu_{\mathrm{eff}}+\sigma_{\mathrm{eff}}^2}\left(e^{\sigma_{\mathrm{eff}}^2}-1\right) \\
\mathrm{Var_L^{(l)}} &= (l - n_{\mathrm{chrom}})e^{2\mu_{\mathrm{eff}}+\sigma_{\mathrm{eff}}^2}\left(e^{\sigma_{\mathrm{eff}}^2}-1\right).
\end{aligned}
\tag{6.54}
$$

With this, approximate $\hat{p}(y|i_1,0)$ by a weighted sum of lognormal distributions $\forall i_1 \geq 1$:

$$
\tilde{p}(y|i_1,0) = \sum_{l=0}^{2n_{\mathrm{chrom}}} p_\mathrm{L}(l|i_1,0)\log\mathcal{N}\left(y|\tilde{\mu}^{(l)},(\tilde{\sigma}^{(l)})^2\right),
\tag{6.55}
$$

with the parameters $\tilde{\mu}^{(l)}$ and $\tilde{\sigma}^{(l)}$ as given in (6.53).

Divisions during delabeling   After divisions during delabeling, using (6.38) and (6.46), the measured label densities are $\forall i_1 \geq 1, i_2 \geq 1$:

$$
\begin{aligned}
\hat{p}(y|i_1,i_2) &= \int_{\mathbb{R}_+} p(\zeta|i_1,i_2)\log\mathcal{N}(y-\zeta|\mu_\mathrm{a},\sigma_\mathrm{a}^2)d\zeta \\
&= \int_{\mathbb{R}_+} \sum_{l=0}^{2n_{\mathrm{chrom}}} \sum_{u=0}^{2n_{\mathrm{chrom}}} \left(p_\mathrm{L}(l|i_1,i_2)\,p_\mathrm{U}(u|i_2)\int_{\mathbb{R}_+} \log\mathcal{N}(\zeta-\chi|\mu_{\mathrm{eff}}+\log(u),\sigma_{\mathrm{eff}}^2) \right. \\
&\qquad \left. \cdot \mathrm{conv}_{j=1}^{l-u}\log\mathcal{N}\left(\chi|\mu_{\mathrm{eff}},\sigma_{\mathrm{eff}}^2\right)\right)d\chi\;\log\mathcal{N}(y-\zeta|\mu_\mathrm{a},\sigma_\mathrm{a}^2)d\zeta \\
&= \sum_{l=0}^{2n_{\mathrm{chrom}}} \sum_{u=0}^{2n_{\mathrm{chrom}}} p_\mathrm{L}(l|i_1,i_2)\,p_\mathrm{U}(u|i_2)\int_{\mathbb{R}_+}\int_{\mathbb{R}_+} \log\mathcal{N}(\zeta-\chi|\mu_{\mathrm{eff}}+\log(u),\sigma_{\mathrm{eff}}^2) \\
&\qquad \cdot \mathrm{conv}_{j=1}^{l-u}\log\mathcal{N}\left(\chi|\mu_{\mathrm{eff}},\sigma_{\mathrm{eff}}^2\right)d\chi\;\log\mathcal{N}(y-\zeta|\mu_\mathrm{a},\sigma_\mathrm{a}^2)d\zeta.
\end{aligned}
\tag{6.56}
$$

Again, the nested convolutions of lognormal distributions, as they appear in each summand, can be replaced by one approximating lognormal distribution according to Fenton's approach (Fenton, 1960),

$$
\begin{aligned}
&\int_{\mathbb{R}_+}\int_{\mathbb{R}_+} \log\mathcal{N}(\zeta-\chi|\mu_{\mathrm{eff}}+\log(u),\sigma_{\mathrm{eff}}^2) \\
&\qquad \cdot \mathrm{conv}_{j=1}^{l-u}\log\mathcal{N}\left(\chi|\mu_{\mathrm{eff}},\sigma_{\mathrm{eff}}^2\right)d\chi\;\log\mathcal{N}(y-\zeta|\mu_\mathrm{a},\sigma_\mathrm{a}^2)d\zeta \\
&\approx \log\mathcal{N}\left(y|\tilde{\mu}^{(l,u)},(\tilde{\sigma}^{(l,u)})^2\right),
\end{aligned}
\tag{6.57}
$$

where the parameters of the approximating lognormal distribution are

$$
\begin{aligned}
\tilde{\mu}^{(l,u)} &= \log\left(\mathrm{E_a} + \mathrm{E_C^{(u)}} + \mathrm{E_L^{(l,u)}}\right) - \frac{1}{2}\log\left(\frac{\mathrm{Var_a} + \mathrm{Var_C^{(u)}} + \mathrm{Var_L^{(l,u)}}}{(\mathrm{E_a} + \mathrm{E_C^{(u)}} + \mathrm{E_L^{(l,u)}})^2} + 1\right) \\
\tilde{\sigma}^{(l,u)} &= \sqrt{\log\left(\frac{\mathrm{Var_a} + \mathrm{Var_C^{(u)}} + \mathrm{Var_L^{(l,u)}}}{(\mathrm{E_a} + \mathrm{E_C^{(u)}} + \mathrm{E_L^{(l,u)}})^2} + 1\right)}.
\end{aligned}
\tag{6.58}
$$

To obtain this, the first two moments of the individual distributions are used which give

$$
\begin{aligned}
\mathrm{E}_a &= e^{\mu_a + \frac{\sigma_a^2}{2}} \\
\mathrm{E}_\mathrm{C}^{(u)} &= u e^{\mu_\mathrm{eff} + \frac{\sigma_\mathrm{eff}^2}{2}} \\
\mathrm{E}_\mathrm{L}^{(l,u)} &= (l-u) e^{\mu_\mathrm{eff} + \frac{\sigma_\mathrm{eff}^2}{2}} \\
\mathrm{Var}_a &= e^{2\mu_a + \sigma_a^2} \left(e^{\sigma_a^2} - 1 \right) \\
\mathrm{Var}_\mathrm{C}^{(u)} &= u^2 e^{2\mu_\mathrm{eff} + \sigma_\mathrm{eff}^2} \left(e^{\sigma_\mathrm{eff}^2} - 1 \right) \\
\mathrm{Var}_\mathrm{L}^{(l,u)} &= (l-u) e^{2\mu_\mathrm{eff} + \sigma_\mathrm{eff}^2} \left(e^{\sigma_\mathrm{eff}^2} - 1 \right).
\end{aligned}
\tag{6.59}
$$

With this, $\hat{p}(y|i_1, i_2)$ can be approximated by a weighted sum of lognormal distributions $\forall i_1 \geq 1, i_2 \geq 1$:

$$
\tilde{p}(y|i_1, i_2) = \sum_{l=0}^{2n_\mathrm{chrom}} \sum_{u=0}^{2n_\mathrm{chrom}} p_\mathrm{L}(l|i_1, i_2) \, p_\mathrm{U}(u|i_2) \log \mathcal{N}\left(y | \tilde{\mu}^{(l,u)}, (\tilde{\sigma}^{(l,u)})^2 \right),
\tag{6.60}
$$

with the parameters $\tilde{\mu}^{(l,i_1)}$ and $\tilde{\sigma}^{(l,i_1)}$ as given in (6.58), and the probability weights $p_\mathrm{L}(l|i_1, i_2)$ as in (6.23), $p_\mathrm{U}(u|i_2)$ as in (6.34).

A further simplification was found which supersedes the summation over u and hence also the calculation of $p_\mathrm{U}(u|i_2)$: Instead of summing over all possible values of u, we replace the values u by the expectation value of the proportion of strands originating from the last uplabeling division event, $\mathbb{E}(U/L)$, multiplied by the number of labeled strands, l:

$$
u \approx \mathbb{E}\left(\frac{U}{L} \right) l = \mathbb{E}\left(\frac{U(i_1, i_2)}{L(i_1, i_2)} \right) l.
\tag{6.61}
$$

The motivation behind this is that the probability of inheriting a labeled strand is independent from the fact whether it originates from the last division, or from an earlier division during uplabeling. Because of this also, the expectation value of the proportion, $\mathbb{E}(U/L)$, is independent of the number of undergone divisions during delabeling: $\mathbb{E}(U(i_1, i_2)/L(i_1, i_2)) = \mathbb{E}(U(i_1, 0)/L(i_1, 0))$. Using furthermore that $U(i_1, 0) = n_\mathrm{chrom}$, and the expression $L(i_1, 0)/2n_\mathrm{chrom} = 1 - 2^{-i_1}$ (Ganusov and De Boer, 2013), the expectation value simplifies to

$$
\mathbb{E}\left(\frac{U}{L} \right) = \begin{cases} 0 & , i_1 = 0, \\ \mathbb{E}\left(\frac{U(i_1, i_2)}{L(i_1, i_2)} \right) = \mathbb{E}\left(\frac{n_\mathrm{chrom}}{L(i_1, 0)} \right) = \frac{n_\mathrm{chrom}}{\mathbb{E}(L(i_1, 0))} = \frac{n_\mathrm{chrom}}{2n_\mathrm{chrom}(1 - 2^{-i_1})} = \frac{1}{2 - 2^{-i_1 + 1}} & , i_1 \geq 1 \end{cases}.
\tag{6.62}
$$

With this, $\hat{p}(y|i_1, i_2)$ can be approximated by a simpler weighted sum of lognormal distributions $\forall i_1 \geq 1, i_2 \geq 1$:

$$
\tilde{p}(y|i_1, i_2) = \sum_{l=0}^{2n_\mathrm{chrom}} p_\mathrm{L}(l|i_1, i_2) \log \mathcal{N}\left(y | \tilde{\tilde{\mu}}^{(l,i_1)}, (\tilde{\tilde{\sigma}}^{(l,i_1)})^2 \right),
\tag{6.63}
$$

with the parameters $\tilde{\tilde{\mu}}^{(l,i_1)}$ and $\tilde{\tilde{\sigma}}^{(l,i_1)}$ similar to $\tilde{\mu}^{(l,i_1)}$ and $\tilde{\sigma}^{(l,i_1)}$ as in (6.58), but with the following moments replaced by

$$\mathrm{E}_{\mathrm{C}}^{(u)} \approx \mathrm{E}_{\mathrm{C}}^{(l,i_1)} = \mathbb{E}\left(\frac{U}{L}\right) l \, e^{\mu_{\mathrm{eff}} + \frac{\sigma_{\mathrm{eff}}^2}{2}}$$

$$\mathrm{E}_{\mathrm{L}}^{(l,u)} \approx \mathrm{E}_{\mathrm{L}}^{(l,i_1)} = \left(1 - \mathbb{E}\left(\frac{U}{L}\right)\right) l \, e^{\mu_{\mathrm{eff}} + \frac{\sigma_{\mathrm{eff}}^2}{2}}$$

$$\mathrm{Var}_{\mathrm{C}}^{(u)} \approx \mathrm{Var}_{\mathrm{C}}^{(l,i_1)} = \left(\mathbb{E}\left(\frac{U}{L}\right) l\right)^2 e^{2\mu_{\mathrm{eff}} + \sigma_{\mathrm{eff}}^2} \left(e^{\sigma_{\mathrm{eff}}^2} - 1\right) \qquad (6.64)$$

$$\mathrm{Var}_{\mathrm{L}}^{(l,u)} \approx \mathrm{Var}_{\mathrm{L}}^{(l,i_1)} = \left(1 - \mathbb{E}\left(\frac{U}{L}\right)\right) l \, e^{2\mu_{\mathrm{eff}} + \sigma_{\mathrm{eff}}^2} \left(e^{\sigma_{\mathrm{eff}}^2} - 1\right),$$

with $\mathbb{E}(U/L)$ as in (6.62).

We evaluated the goodness of the approximations (6.55) and (6.63), obtained by imposing the assumption (A1) as well as the Fenton approximation for (6.55) and the Fenton-like approach for (6.63). Therefore, we implemented the process of label uptake and inheritance as described by (P1)–(P5), with a labeling efficacy sampled for each individual division event from a lognormal distribution, and an autofluorescence also sampled from a lognormal distribution. We then ran stochastic simulations of 10000 dividing cells under this process. The obtained histograms of the resulting label distribution after (i_1, i_2) divisions were normalized and then compared to the distributions obtained by evaluating (6.55) and (6.63). Figure 6.5 shows the results for the example of human cells with $n_{\mathrm{chrom}} = 46$, a labeling efficacy with $\mu_{\mathrm{eff}} = 1, \sigma_{\mathrm{eff}} = 0.8$, and an autofluorescence with $\mu_{\mathrm{a}} = 3, \sigma_{\mathrm{a}} = 0.5$, for division numbers $i_1 = 1, \ldots, 6$ and $i_2 = 0, \ldots, 3$. As can be seen in Figure 6.5, the approximation is highly accurate, which was also found for further division numbers and other labeling parameters.

The distributions (6.50), (6.55), and (6.63) provide expressions for the measured label densities which can be evaluated with low computational effort, since they merely represent weighted sums of lognormal distributions. While the numerical evaluation of the convolutions appearing in (6.51) and (6.56) takes several minutes, the calculation of the approximating distributions (6.55) and (6.63) can be conducted in less than a second, and is thereby orders of magnitudes faster. At the same time, the approximation is of high accuracy, as was seen in Figure 6.5.

Finally, the initial task of efficiently calculating the measured overall label distribution can be solved. The individual measured label distributions can be derived via $\hat{n}(y, \mathbf{i}, j|t) = N(\mathbf{i}, j|t)\hat{p}(y|\mathbf{i})$, and from this in turn the measured overall label distribution $\hat{m}(y|t)$ as in (6.48) or, to replace the infinite sums, a truncated version of the measured overall label distribution

$$\hat{m}_S(y|t) = \sum_{i_1=0}^{S-1} \sum_{i_2=0}^{S-1} \sum_{j=1}^{J} \hat{n}(y, (i_1, i_2), j|t). \qquad (6.65)$$

This is the output of the model which corresponds to the measurement output, and thus is the variable that should be fit to measurement data. We will demonstrate in the following section in an example how parameter estimation, using our model and data from BrdU labeling experiments, serves to identify proliferation parameters in a multi-cell type population.

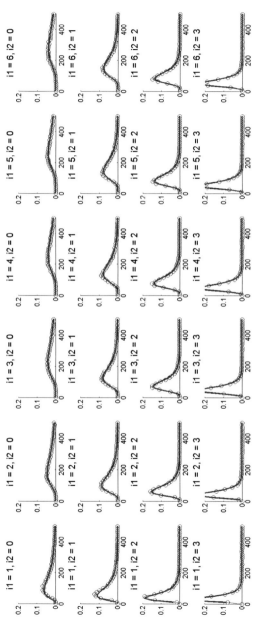

Figure 6.5: Evaluation of the approximation goodness. Approximated distribution (red line) versus normalized histograms from stochastic simulations (blue circles ○). For the approximation and the stochastic simulations, we used $n_{chrom} = 46$ as in human cells, a labeling efficacy with $\mu_{eff} = 1$, $\sigma_{eff} = 0.8$, and an autofluorescence with $\mu_a = 3$, $\sigma_a = 0.5$. The stochastic simulations were conducted for 10000 cells, that take up label into newly built strands, and inherit labeled strands, as described in the text by (P1)–(P5).

6.3 Examples: Model-based studies on BrdU labeling

In this section, we employ the presented model for two simulation studies: In the first one, our model serves to investigate potential sources for the heterogeneity observed in BrdU labeling data, and thus to promote a better understanding of the processes involved in the labeling procedure. In the second study, we employ our model along with existing models to estimate proliferation parameters from BrdU data, wherewith it will be seen that our model provides more accurate and less uncertain results.

6.3.1 Potential sources of heterogeneity in BrdU data

Data from BrdU labeling experiments, for example the work of Takizawa et al. (2011), show that labeled cells exhibit quite heterogeneous BrdU label intensities, in the sense that the measured label intensities are spread out over values of several magnitudes. This holds even true for cells that have undergone the same number of divisions, as Takizawa et al. (2011) made impressively visible: They labeled cells with BrdU and jointly with CFSE, such that the latter allows to conclude about the maximum number of undergone divisions during BrdU uplabeling. Up to date, it is not fully understood why BrdU-labeled cells exhibit such heterogeneous label intensities, which prohibits to directly conclude from a cell's label intensity to its division number. In this subsection, we exploit our model to investigate and discuss several potential causations for this heterogeneity.

As we have seen, the process of *DNA label segregation* is partly deterministic and partly stochastic, as the inheritance of labeled DNA depends on the segregation of chromosomes. The mathematically rigorous model for DNA label segregation that we have derived in Subsection 6.2.2 allows to reason about the influence of chromosome segregation on the DNA label distribution. It was seen that DNA label segregation induces heterogeneity in the number of labeled strands for division numbers $i_1 \geq 2$, but importantly not for $i_1 = 1$. For cells that have undergone only one division during uplabeling, their number of labeled strands is determined to equal the number of chromosomes, n_{chrom}, thus is the same among all cells of division number $i_1 = 1$. This, however, contradicts the observation (Takizawa et al., 2011) that cells already of division numbers $i_1 = 0, 1$ show highly heterogeneous BrdU intensities. Thus we can conclude that DNA label segregation on its own can not be responsible for the heterogeneity in BrdU data.

Besides the segregation of chromosomes (and thus DNA label), also the *different sizes of chromosomes* were hypothesized as a possible causation for the observed label heterogeneity. However, we can rule out that this is the major cause for label heterogeneity by the following argumentation: Let us denote the size of the ι-th chromosome by $S_\iota, \iota = \{1, \ldots, n_{chrom}\}$. Each cell then contains the same set of chromosome sizes, $\{S_1, \ldots, S_{n_{chrom}}\}$. Considering strands as the carrier units of DNA label, each cell is known to contain two strands of each size, $\{S_1, S_1, \ldots, S_{n_{chrom}}, S_{n_{chrom}}\}$. After having undergone one uplabeling division, $i_1 = 1$, each ι-th chromosome consists of exactly one unlabeled strand of size S_ι and one labeled strand of size S_ι. That means that a cell with division number $i_1 = 1$, regardless of the differences in chromosome sizes, has a purely deterministic compilation of labeled strands, namely $\sum_{\iota=1}^{n_{chrom}} S_\iota$. Thus, all cells of division number $i_1 = 1$ would still exhibit the same summarized label content. Again, this is in contradiction to the heterogeneity in BrdU intensities for division number $i_1 = 1$ (Takizawa et al., 2011). Although the differences in chromosome sizes can be expected to influence the variability in label intensities for higher division numbers $i_1 \geq 2$, it can not explain the heterogeneity for $i_1 = 1$, and thus not be the main source for heterogeneous label intensities.

As it is seen that neither chromosome segregation nor differences in chromosome size can fully explain the heterogeneity in BrdU label intensities, we next discuss the effects of a heterogeneous labeling efficacy and of the autofluorescence. Both factors are covered by the model that was developed above. To illustrate and quantify these effects, we simulated the label intensity distributions of a proliferating cell population for which we assume that there is no heterogeneity from autofluorescence, or no heterogeneity from labeling efficacy, respectively. For this cell population, we assumed $n_{\text{chrom}} = 46$, which is the number of chromosomes in human cells. Furthermore, we used for simplicity one cell type, $J = 1$, $N_{\text{ini}} = 1000$ cells, and proliferation parameters $\forall i \in \mathbb{N}_0^2 : \alpha_i = \beta_i = 0.05[1/d]$ which render the population size at steady state. The labeling parameters were chosen to be $\mu_{\text{eff}} = 1, \mu_a = 3$, whereas one of σ_{eff} and σ_a was set to 0 while the other was set to 0.5, depending on the respective scenario to be investigated. The uplabeling phase was up to day $T_1 = 21$, followed by a delabeling phase until day $T_2 = 70$. We considered $S = 20$ division numbers which yields a highly accurate model output with a truncation error (4.23) of less than 10^{-4}, and assumed that the measurement is based on 256 fluorescence intensity channels.

By considering a heterogeneous *labeling efficacy* we reflect in the model that the uptaken label may not be a fixed value, but more generally may vary between each label incorporation event. In contrast to the just discussed factors of DNA label segregation and chromosome sizes, the labeling efficacy can cause heterogeneity also for the first division number $i_1 = 1$, as is visible in the label densities (6.31) even though the number of labeled strands is a deterministic value, giving for $i_1 = 1$: $p_L(n_{\text{chrom}}|1, 0) = 1$ and $p_L(l \neq n_{\text{chrom}}|1, 0) = 0$. We illustrate the heterogeneity in measured label intensity induced by labeling efficacy alone via the simulation setting described above with $\sigma_{\text{eff}} = 0.5$, $\sigma_a = 0$. A labeling efficacy with sufficiently large standard deviation leads to a broadening of the label densities, such that the peaks of different division numbers can hardly be separated any more, as seen in Figure 6.6. Thus we can conclude that a sufficiently large labeling efficacy is sufficient to induce heterogeneity in measured label intensities.

The *autofluorescence* in turn is also capable of generating a certain heterogeneity in label intensities. Again, we quantify this effect via the simulation setting described above, now with $\sigma_{\text{eff}} = 0$ and $\sigma_a = 0.5$. As can be seen in Figure 6.7, an autofluorescence with sufficiently large standard deviation induces heterogeneity in the cells with low label intensity, that is with low division numbers during uplabeling or/and high division numbers during delabeling. Importantly, at higher label intensities (corresponding to higher division numbers during uplabeling, and low division numbers during delabeling), the impact of the autofluorescence vanishes, and the individual peaks corresponding to different division numbers become again visible. Thus the autofluorescence, even with a large standard deviation, can not be the only causation of heterogeneity in label intensities.

From these investigations, we can conclude that neither chromosome segregation, nor chromosome size differences, nor the autofluorescence alone can be the only source for the heterogeneity observed in BrdU label intensities. In contrast, a heterogeneous label uptake as reflected in the labeling efficacy can induce heterogeneity in BrdU label intensities, even if the other factors were nonheterogeneous. However, one may argue that also a combination of

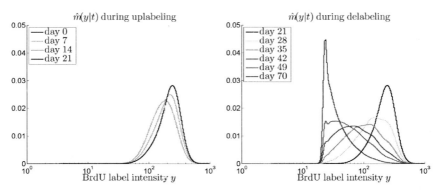

Figure 6.6: Heterogeneity induced by labeling efficacy alone, by setting $\sigma_a = 0$. It can be seen that a labeling efficacy with sufficiently large standard deviation (here: $\sigma_{eff} = 0.5$) already on its own leads to broadening of the peaks such that they can hardly be separated.

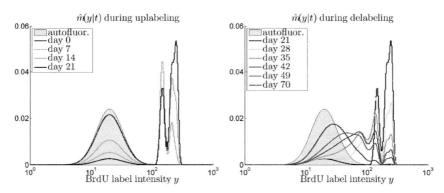

Figure 6.7: Heterogeneity induced by autofluorescence alone, by setting $\sigma_{eff} = 0$. It can be seen that the autofluorescence, even with a large standard deviation (here: $\sigma_a = 0.5$), can not induce the experimentally observed magnitude of heterogeneity in the division peaks.

chromosome segregation (generating heterogeneity in higher division numbers) and autofluorescence (creating heterogeneity mainly in undivided cells) could serve as an explanation for heterogeneous BrdU measurements, without the need for an additional heterogeneity source in labeling efficacy. In addition, the measurement by flow cytometry may induce additional measurement noise, for which up to date no exhaustive studies are available, but for which the effect can be expected to be smaller than that observed in the data. Besides the biological plausibility for a labeling efficacy that differs between individual label uptake events, we have seen that it is the only one of the discussed sources that is capable on its own to generate heterogeneity in label intensities for cells of all division numbers. This may further emphasize the relevance of considering a general labeling efficacy for modeling BrdU-labeled populations and analyzing data from BrdU labeling experiments.

6.3.2 Estimation of proliferation parameters from BrdU data

The eventual goal of a model for BrdU-labeled cell populations should be to match it to data, and thus to employ it for inferring the proliferation parameters of interest from BrdU data sets. To evaluate the capability and performance of our model, the aim of this subsection is

- to employ our model for estimating proliferation parameters from typical BrdU data and evaluate its performance, and

- to compare the results obtained with our model to results obtained with models that exploit only the threshold value instead of the full intensity distribution.

For this purpose, we use a comprehensive set of synthetically generated data, which offers several advantages for the evaluation of the models' performances: Firstly, an obvious advantage is that the true parameters are known, and thus the models can be evaluated with respect to their capability and accuracy in recovering the true parameters. Secondly, synthetic data can be generated to an amount that is practically not realizable by wet lab experiments. A comprehensive data set in contrast allows to perform also statistics on the estimation results. Thirdly, additional noise sources such as measurement noise are excluded. Their suitable treatment requires sophisticated analysis and modeling for its own, which is out of the scope of this work. Insufficient knowledge about measurement noise distorts the parameter estimation and thus the comparison between different models.

Data setup We generated synthetic data by implementing and simulating the DNA label segregation process as subsumed by (P1)–(P5), with settings in the style of typical BrdU labeling experiments (Mohri et al., 1998; Takizawa et al., 2011) as pointed out in the following. The uplabeling phase was defined to last 21 days, thus $(0, T_1] = (0, 21]$, followed by a delabeling phase $(T_1, T_2] = (21, 70]$. The autofluorescence (6.46) was set with $\mu_a = 3, \sigma_a = 0.5$, and a labeling efficacy (6.25) with $\mu_{\mathrm{eff}} = 1, \sigma_{\mathrm{eff}} = 0.8$. The values of these labeling parameters were chosen based on visual inspection of the BrdU intensities shown in the experiments by Takizawa et al. (2011).

The cell population in these virtual experiments was assumed to consist of $N_{\mathrm{ini}} = 10000$ human cells with $n_{\mathrm{chrom}} = 46$ chromosomes, comprising two cell types $J = 2$, with equal subpopulation sizes: $N_{\mathrm{ini},1} = N_{\mathrm{ini},2} = 0.5 N_{\mathrm{ini}}$. For simplicity of the setting, we assumed that there is no exchange between the cell types, thus $\forall \mathbf{i} \in \mathbb{N}_0^2, j_1, j_2 \in \{1, 2\}, \forall t : \delta_{\mathbf{i}}^{j_2 j_1}(t) = 0$, and $\forall j_2 \neq j_1 : w_{\mathbf{i}}^{j_2 j_1}(t) = 0; w_{\mathbf{i}}^{j_1 j_1}(t) = 1$. We chose division and death rates to be cell type-dependent, $\forall \mathbf{i} \in \mathbb{N}_0^2, j \in \{1, 2\}, \forall t : \alpha_{\mathbf{i}}^j(t) = \alpha^j, \beta_{\mathbf{i}}^j(t) = \beta^j$, and furthermore both cell type subpopulations to be at steady state, $\forall j \in \{1, 2\} : \beta^j = \alpha^j$. One cell type was chosen to proliferate slowly, $\alpha^1 = 0.01[1/d]$, and the other cell type to proliferate fast, $\alpha^2 = 0.2[1/d]$.

Measurements were taken for 50 samples, each comprising independent replicates at nine time points, days $t \in \{0, 7, 14, 21, 28, 35, 42, 49, 70\}$. The measurements were obtained as binned snapshot data with 256 intensity bins, ranging from 10^0 to 10^4. The corresponding $9 \cdot 50 = 450$ data sets, each with 10000 cells, were obtained by running the stochastic simulation of (P1)–(P5) $9 \cdot 50 \cdot 10000 = 4.5 \cdot 10^6$ times, whereas each cell's resulting label intensity was tracked up to division numbers $(i_1, i_2) = ((S-1), (S-1)) = (39, 39)$. This yielded histograms of label intensities of cells that have divided (i_1, i_2)-times, corresponding to an experimental assessment of the label densities (6.55), (6.60), $\tilde{p}(y|i_1, i_2)$. To assess the measured overall label distribution (6.65), $\hat{m}_S(y|t)$, we also stochastically simulated the number of cells that have divided (i_1, i_2)-times, $N(i_1, i_2|t)$, for each measured time point t. With this, 50 independent data sets, each comprising 9 time points with independent replicates, of the measured overall label distribution were achieved.

Model setup The models that were fit to the data were all based on the model equations (6.41), truncated with $S = 40$ division numbers. Based on that, we derived eight variants of models and outputs that we fit to the data and compared the results: In order to assess the performance of models with different information content, we used four output variants. In addition, we used model variants that assumed two cell types as well as model variants assuming one cell type, in order to evaluate whether the presence of two cell types was correctly identified, or whether a two-celltypes model might overfit the data. All models used the same parameter values as indicated in the description of the data setup, except for the free parameters to be estimated, as pointed out below.

Regarding the model output, the following four alternatives were considered: Models denoted by "-hist" gave the full intensity distribution $\hat{m}_S(y|t)$. Models denoted by "-perc-<percentile>" gave the percentage of BrdU positive cells, $|\{l^\nu | l^\nu \geq l_\theta\}|/|\{l^\nu\}|$, in dependence of the threshold value l_θ. To assess the impact of the choice of the threshold value on the parameter estimation results, we used three different values l_θ that were defined based on the $90\%, 97\%, 99\%$-percentile of unlabeled (undivided) cells lying below the threshold value l_θ. Regarding the number of cell types, the two alternatives were as follows: "M1" denotes models assuming one cell type, thus having three parameters to be estimated, namely the proliferation rate α, and the labeling parameters $\mu_{\text{eff}}, \sigma_{\text{eff}}$. "M2" denotes models that consider two cell types, thus have five parameters to be estimated, namely the proliferation rates α^1, α^2 of the two cell types, the fraction of cell types $f_1 := N_{\text{ini},1}/N_{\text{ini}}$, and the labeling parameters $\mu_{\text{eff}}, \sigma_{\text{eff}}$.

With this, eight models were defined and compared: (M1-hist), (M1-perc-90%), (M1-perc-97%), (M1-perc-99%), (M2-hist), (M2-perc-90%), (M2-perc-97%), (M2-perc-99%). Each of these models was employed for parameter estimation, which was conducted by a multi-start optimization approach with 10 starts each. Only the start with the best fit (lowest objective function value) was used as a result. Further details of the parameter estimation procedure will be made available by Schittler et al. (in prep.).

Results The difference in the optimal objective function values between models assuming one cell type (M1-...) versus models assuming two cell types (M2-...) was always orders of magnitudes higher for the M1-models than for the according M2-models. This means that, regardless of the used output, all models correctly favored the hypothesis that two distinct cell types are present in the population. Exemplary fits of the models using the full intensity distribution are shown in Figure 6.8 for the two-celltypes model (M2-hist), and in Figure 6.9 for the one-celltype model (M1-hist). For comparison, also an exemplary fit of a model with two cell types but using only the threshold value of the 97%-percentile (M2-perc-97%) is shown in in Figure 6.10. As can be seen in these figures, the model assuming two cell types (Figure 6.8) fits the data clearly better than the model neglecting the existence of cell types with different proliferation properties (Figure 6.9), and also than the model using only the threshold value (Figure 6.10). This emphasizes once again the need for models accounting for cell types as well as for the full measurement output, if data from BrdU experiments should be reproduced accurately.

The estimation results of fitting the two-celltypes models (M2-hist), (M2-perc-90%), (M2-perc-97%), (M2-perc-99%) to the data are summarized in Figure 6.11. From there it is seen that the model using the full intensity profile, (M2-hist), provides more accurate and less uncertain parameter estimation results. Furthermore, it does not depend on the choice of the threshold value. In comparison, the estimation results of fitting the one-celltype models (M1-hist), (M1-perc-90%), (M1-perc-97%), (M1-perc-99%) to the data are summarized in Figure 6.12. There it is seen that the estimated parameter values are more spread out, thus no uniquely determined parameter region could be found that fits the data best. Interestingly, the model using the full intensity distribution (M1-hist) identifies two parameter regions, one with a high and the other with a low proliferation rate α. The high proliferation rate is close to the value of the fast proliferating cell type, whereas the low proliferation rate is close to the one of the slow proliferating cell type. Thus, the results obtained by using the full model output may serve as an indicator that two cell types with clearly distinct proliferation properties are present in the population, which in contrast is not pronounced in the results using the threshold output, (M1-perc-...).

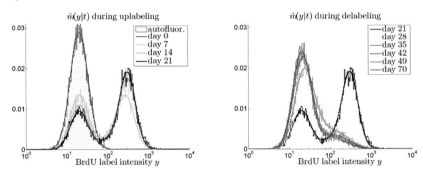

Figure 6.8: Exemplary fit of the model assuming two cell types and using the full intensity profile (M2-hist) to a data set. The estimated parameter values were: $\alpha^1 = 0.0103, \alpha^2 = 0.2005, f_1 = 0.5018, \mu_{\text{eff}} = 0.9552, \sigma_{\text{eff}} = 0.7936$. Simulation output is shown in thick lines, data is shown in thin lines.

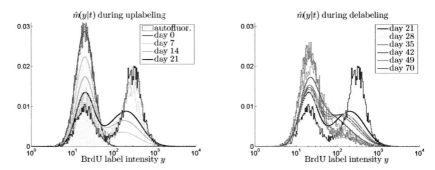

Figure 6.9: Exemplary fit of the model assuming one cell type and using the full intensity profile (M1-hist) to the same data set as in Figure 6.8. The estimated parameter values were: $\alpha = 0.0191, \mu_{\text{eff}} = 0.9636, \sigma_{\text{eff}} = 1$. Simulation output is shown in thick lines, data is shown in thin lines.

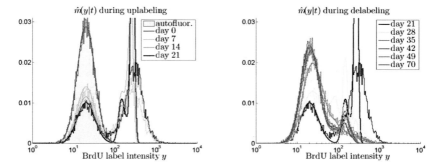

Figure 6.10: Exemplary fit of the model assuming two cell types and using the threshold value of the 97%-percentile (M2-perc-97%) to the same data set as in Figure 6.8. The estimated parameter values were: $\alpha^1 = 0.0090, \alpha^2 = 0.1845, f_1 = 0.5116, \mu_{\text{eff}} = 1.0051, \sigma_{\text{eff}} = 0.1849$. Simulation output is shown in thick lines, data is shown in thin lines.

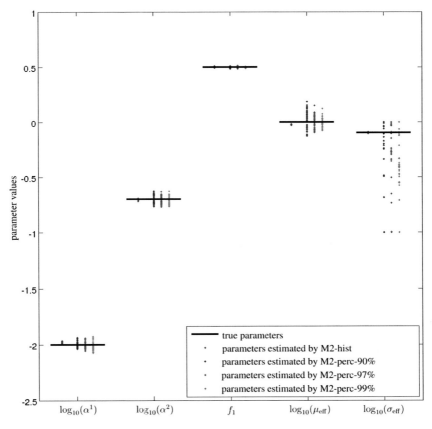

Figure 6.11: Results of the parameter estimation, by fitting the models with two cell types to 50 data sets, as described in the text. Estimated parameters were the proliferation rates α^1 and α^2, the cell type fraction, f_1, as well as the labeling efficacy parameters μ_{eff} and σ_{eff}. The true parameters are denoted by black lines. The estimated parameter values from the best fit for each of the 50 data sets by using four different model outputs: the full intensity distribution (M2-hist; blue dots), or the percentage of BrdU positive cells (M2-perc-90%,-97%,-99%; dark red, red, and orange dots).

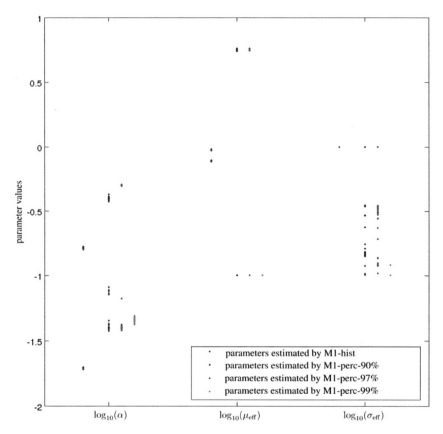

Figure 6.12: Results of the parameter estimation, by fitting the models with one cell type to 50 data sets, as described in the text. Estimated parameters were the proliferation rate α, as well as the labeling efficacy parameters μ_{eff} and σ_{eff}. The estimated parameter values from the best fit for each of the 50 data sets by using four different model outputs: the full intensity distribution (M1-hist; blue dots), or the percentage of BrdU positive cells (M1-perc-90%,-97%,-99%; dark red, red, and orange dots).

6.4 Summary and discussion

In this chapter, we developed a model for BrdU-labeled multi-cell type populations. For this purpose, we first derived a mathematically rigorous description of the DNA label segregation process (Problem 6.1), and then addressed the overall problem of modeling BrdU-labeled multi-cell type populations (Problem 6.2). The presented model accounts for division numbers during uplabeling and delabling, for cell types and cell type-specific proliferation parameters, and for the full BrdU intensity distribution. To the best of our knowledge, this is the first mathematically rigorous treatment of DNA label segregation, as well as the first model for BrdU labeling that can exploit the full information from data based on a mechanistic description of the label intensity.

As it was seen, the mathematical description of DNA label segregation is nontrivial. The reason is that the segregation of labeled DNA strands is partly deterministic, partly stochastic. The mathematical model that we have derived here gives the probability distribution for the number of labeled strands in dependence of undergone divisions. Already this first step towards a comprehensive model for BrdU-labeled cell populations yields new insight into some characteristics of BrdU labeling: We found that the number of labeled strands after one division during uplabeling is a deterministic value, namely half of the number of chromosomes, as to be expected. However, for all higher division numbers as well as division numbers during delabeling, the probability distributions for the number of labeled strands are given by weighted sums of binomial distributions.

To derive a mathematical description of the BrdU label concentration in cells, dependent on their division number, we combined the derived model for DNA label segregation with an effective label uptake, given by a labeling efficacy distribution. As we discussed, its exact derivation would require to consider all possible combinations of labeling events in the division history of cells, which would result in an intractably complex model. Fortunately, we could show that this can be circumvented by imposing two simplifying assumptions, resulting in an efficiently computable approximation that at the same time is highly accurate. With these results and the properties of BrdU labeling, we arrived at a DsCL model which can be solved by a decomposition approach. This not only enables an efficient solution scheme, but also represents the first mathematically rigorous model for the BrdU concentration of proliferating cells, which furthermore can be of several cell types.

In order to provide a link of our model to data from BrdU experiments, in a further step we also incorporated the autofluorescence, which together with the beforehand derived BrdU label concentration provides the measured label distribution. To calculate the measured label distribution, we presented an efficient approach based on moment matching. This results in a highly accurate approximation which is given by a sum of lognormal distributions, and can thus be computed highly efficiently.

We presented two studies in which we demonstrate how our model can serve to analyze BrdU data: In the first study, we investigated potential sources for heterogeneity in BrdU data. For example, it could be seen that neither DNA label segregation, nor different sizes of chromosomes can fully account for the observed heterogeneities. In contrast, the here introduced labeling efficacy can, which suggests that heterogeneity in BrdU intensities may originate mainly from heterogeneous label uptake. In a second study, we employed our model to estimate proliferation parameters from BrdU data. Using a comprehensive set of synthetic data, we showed that our model provides more accurate and less uncertain results than existing models. In this way, our model promotes the assessment of proliferation properties from BrdU data, including the detection of multiple cell types differing in their proliferation rates within one population.

One may argue that as the here presented model provides a more complex output, it comes at a higher computational cost than existing models (Bonhoeffer et al., 2000; De Boer et al., 2003b,c; Kiel et al., 2007; Mohri et al., 1998; Ganusov and De Boer, 2013; Glauche et al., 2009; Parretta et al., 2008; Wilson et al., 2008). This is indeed correct, however, we have shown that the derivation of the label distribution can be achieved by highly efficient and yet highly accurate approximations. Furthermore, the advantage that the full information from experimental data can be used should counterbalance the additional computational burden, as the time, effort and costs for conducting such labeling experiments can be expected to be clearly higher. The advantage of our model against existing models that use less of the data information may be even more pronounced in experimental data where additional measurement noise is present.

Chapter 7

Conclusions

7.1 Summary and discussion

In this thesis, a mathematical modeling framework to simulate and analyze cell type transitions has been presented. The contributions comprise generic gene switch models, detailed gene regulatory network models, as well as cell population models accounting for cell types, division numbers, and label concentration. As pointed out at the beginning in Chapter 1, these different modeling approaches are essential for a quantitative understanding of cell type transitions. We have developed novel modeling methods and model classes that integrate several of these up to date isolated aspects, and enable efficient simulation and analysis schemes.

Troughout this thesis, typical problems arising in the context of cell type transitions have been cast into a mathematical framework, and approaches from diverse disciplines have been adopted to solve these problems. As in systems and control theory, the stability properties of feedback systems have been analyzed via the Nyquist criterion. From system dynamics, bifurcation analysis served to study the impact of parameters on a system's stability properties. From mathematics, solution methods for ODEs and PDEs have been employed. With the help of statistical methods, complex probability distributions have been phrased into computationally treatable expressions.

For a typical modeling problem in the context of cell differentiation, we have demonstrated in Chapter 2 how a gene switch model can be developed systematically, such that the model reproduces the properties of the biological system. We have derived a deterministic model which was then exhaustively studied via multistability and bifurcation analysis, and which furthermore has been employed for simulations of the single cell dynamics. In addition, dynamics on the cell population level have been simulated and analyzed via a stochastic version of the gene switch model. With this, we have shown how mathematical and systems-theoretical tools can assist to develop models which are suited to elucidate the behavior of cell type-determining gene regulatory systems.

We have proposed the obtained gene switch model as a robust and general design principle for gene regulatory determinants in similar differentiating cell systems. Other biological systems that exhibit different cell type transition properties may result in accordingly different models, however, they may be obtained by an analogous procedure of model development and analysis. As a drawback of such core motif models one may argue that they do not account for detailed kinetic mechanisms, and due to their generic character may not be suited to be fit to experimental data. To overcome these shortcomings, generic models as the one presented in

Chapter 2 may be extended to more detailed models by the construction method that we have presented in Chapter 3. With this, the suggested modeling approach for the gene switch may serve as a blueprint for similar problems of modeling gene regulatory systems that determine cell type transitions.

For the purpose of linking core motif models to more detailed gene regulatory network models, the concept of multistability equivalence between gene regulatory network models of different dimensionality has been introduced in Chapter 3. We then have presented a novel construction procedure which was proven to yield a multistability-equivalent system. The benefits of this method have been demonstrated by investigating a gene regulatory network which determines the differentiation of stem cells into three possible mature cell types. The construction procedure allows to derive a more realistic model of gene regulation, for example by incorporating the kinetic details as well as additional gene expression products if it is desired, for example, to fit the model to respective measurements.

The class of gene regulatory network models obtained by this method relies on a modular interaction structure. Such genetic interaction modules have indeed been suggested to be highly prevalent in gene networks (Bar-Yam et al., 2009; Hartwell et al., 1999; Tyson et al., 2003; Zinman et al., 2011). Furthermore, the construction procedure imposes linear activation functions and some mild parameter constraints for the additionally introduced gene expression states. Despite these potential restrictions, our method offers a novel approach to overcoming the gap between models on distinct levels of detail. It thereby opens up new possibilities of integrating the results from multistability analysis into the development and parametrization of more realistic gene regulatory network models.

After having investigated the modeling of cell type transitions on the gene regulatory level, the following chapters were dedicated to modeling and analyzing cell populations which comprise multiple cell types. In Chapter 4, a new and general model class for proliferating multi-cell type populations in labeling experiments has been presented. As we have demonstrated, this enables to develop valuable modeling and solution approaches on the methodological level which hold independent of the specific application. Moreover, the link between cell type-determining gene expression dynamics and multi-cell type population models has been investigated. Although matching the parameters between both model classes relies on further information about the state variables, the interdependency of gene regulatory and cell population dynamics became obvious by deducing necessary conditions for certain dynamical behavior.

In Chapters 5 and 6, two specific models for common labeling techniques have been derived, which benefit from the efficient simulation and analysis schemes of the general multi-cell type model class introduced beforehand. A model tailored to CFSE labeling experiments has been presented in Chapter 5, whereas a model specified for BrdU labeling experiments has been developed in Chapter 6. To the best of the author's knowledge, the proposed models are the first ones that at the same time consider cell types, division numbers, and the respective label dynamics of CFSE or BrdU, and give the full label distribution as an output. Furthermore, the BrdU-specific model incorporates the first mathematically rigorous description of the DNA label segregation process.

In this way, the new models for CFSE- and BrdU-labeled multi-cell type populations are able to exploit the full information from data, and promise to reveal additional information on biologically relevant properties such as cell type specific proliferation dynamics. We have illustrated the capability of our models by simulations of CFSE-labeled multi-cell type populations and model-based studies on typical data from BrdU labeling experiments. It was seen that distinct cell type-specific population properties yield distinct overall label distributions, and the full resolution of label distribution yields more accurate and more reliable estimates of proliferation and subpopulation parameters.

The presented approaches are applicable to a broad range of similar problems which commonly arise in quantitative studies of cell type transitions. We have addressed the modeling and analysis of gene regulatory dynamics which determine cell types, as well as of the dynamics of populations which comprise multiple cell types. In summary, this thesis significantly contributes towards a more holistic mathematical modeling framework for studying cell type transitions.

7.2 Outlook

Although this thesis has made important contributions to the mathematical modeling of cell type transitions, it is clearly out of the scope of one thesis to pursue all questions arising in this context. Several possibilities for future work emerge as will be pointed out in the following.

The aim of the gene switch model as developed in Chapter 2 was to represent the core motif of underlying gene regulatory dynamics for one single cell type transition step within cell differentiation. However, cell differentiation is rather a process of multiple subsequent cell type transition steps which correspond to increasing specialization towards the eventually adopted cell type. To model such a sequence of cell type transitions, interconnections of gene regulatory models somewhat similar to the gene switch model in this thesis have been proposed, for example, by Foster et al. (2009); Zhou et al. (2011). It would be tempting to investigate in a systematic and general context which gene switch motifs and, moreover, which interconnections of gene switches are most suitable for reproducing the cell differentiation processes as observed in biological systems.

In particular, there is an ongoing discussion whether cell differentiation follows a hierarchical pathway of subsequent specializations, or whether several possible cell types are competing on an equal level (Bar-Yam et al., 2009; Chang et al., 2006; Cinquin and Demongeot, 2005; Foster et al., 2009; Huang et al., 2005; Kuchina et al., 2011; MacArthur et al., 2009; Schittler et al., 2010). The answer to this question is essential in order to fully control and successfully achieve the differentiation into a desired cell type. Generic core motif models could serve to compare such alternative hypotheses, and thus to reveal the most plausible explanation for the gene regulatory process of cell differentiation.

The construction method proposed in Chapter 3 provides a novel approach to systematically derive detailed gene regulatory network models from generic core motif models. So far, our method considers merely the preservation of multistability properties. The approach of expanding models to higher dimensions facilitates the transfer of results from systems theoretic analysis to more realistic models, thus it will be worth to also investigate the equivalence of bifurcation properties and vector fields. Besides that, the linear interaction terms currently utilized in the construction procedure restrict the class of obtainable models. Since stability properties are defined via the local linearized dynamics, a generalization to nonlinear interaction terms should be possible, whereas it will require a more elaborated construction procedure and proof of multistability equivalence.

As was seen throughout this thesis, different approaches are suited for modeling the various aspects of gene regulation and cell population dynamics, each of which is important in order to understand cell type transitions. Population balance equation models including gene switch dynamics have been employed to investigate the interdependencies of gene switch dynamics and cell type fluxes, which emerge in population models as discussed in Chapter 4. This study is far from a comprehensive framework, but may serve as a starting point to further investigate and strengthen the connection between these complementary modeling approaches.

The general model class presented in Chapter 4 unifies, to the best of the author's knowledge, for the first time cell types, division numbers, and the full label concentration in one single framework. By accounting for cell type-, division-, and time-dependent proliferation parameters, as well as label degradation, uptake, and inheritance, it allows to deduce more realistic models that can account for biologically relevant processes. However, there may be further aspects such as size structure or age structure and age-dependent proliferation parameters, as for example considered by Gyllenberg (1986); Gyllenberg and Webb (1990); Metzger et al. (2012). Some of these models have already been successfully combined with division numbers or/and cell types. Thus it might be worthwhile to develop an even more general model class incorporating all of the mentioned aspects.

To employ the labeling-specific models for parameter estimation from experimental data, some additional considerations should be noted: In the CFSE-specific population model in Chapter 5, the autofluorescence was not incorporated which was also not necessary for the studies in this work. If required, however, it will be straightforward to exploit a similar approach as for the BrdU-specific model in Chapter 6 and for a related CFSE-specific model by Hasenauer et al. (2012a).

In the BrdU-specific population model in Chapter 6 in turn, we also incorporated the autofluorescence, since we employed this model for estimating parameters from BrdU data. This data was obtained sytetically from stochastic simulations of the full biological process. In contrast to synthetic data, experimental data is in addition commonly corrupted by measurement noise. Therefore, a model's performance in parameter estimation will also highly depend on the derivation of a suitable error model. Here instead, the aim was to evaluate the performance gain if the model considers the full labeling process, in comparison to existing approaches, for which synthetic data as used here may be considered more adequate than experimental data. As the model developed in this work proved to yield more accurate and more reliable parameter estimates, it will be highly interesting to employ our model for estimating parameters from experimental data in future work.

To conclude, the contributions of this thesis open up multiple directions for future work in mathematical modeling of cell types and transitions between them. The thesis at hand provides important steps towards a holistic modeling framework, which will serve a thorough and mathematically found understanding of cell type transitions.

Appendix

Appendix A: Derivation of a dimensionless gene switch model

The original model equations (2.12)–(2.14) can be transformed into the following set of dimensionless model equations:

$$\frac{d}{d\tau_P}\chi_P = \frac{\alpha_P \chi_P^n + 1}{1 + \zeta_D + \gamma_{PP}\chi_P^n} - \chi_P, \tag{A.1}$$

$$\frac{d}{d\tau_O}\chi_O = \frac{\alpha_O \chi_O^n + 1 + \zeta_O}{1 + \gamma_{OO}\chi_O^n + \gamma_{OC}\chi_C^n + \gamma_{OP}\chi_P^n} - \chi_O, \tag{A.2}$$

$$\frac{d}{d\tau_C}\chi_C = \frac{\alpha_C \chi_C^n + 1 + \zeta_C}{1 + \gamma_{CC}\chi_C^n + \gamma_{CO}\chi_O^n + \gamma_{CP}\chi_P^n} - \chi_C, \tag{A.3}$$

This is achieved by introducing dimensionless parameters

$$\begin{aligned}
\alpha_i &= \frac{a_i b_i}{m_i^n k_i^n} & i &\in P, O, C,\\
\gamma_{ii} &= \frac{c_{ii} b_i^n}{m_i^{n+1} k_i^n} & i &\in P, O, C,\\
\gamma_{ij} &= \frac{c_{ij} b_i^n}{m_i m_j^n k_j^n} & i &\in O, C, j \in P, O, C,\\
\zeta_k &= \frac{u_k}{b_i} & (k, i) &\in (D, P), (O, O), (C, C),
\end{aligned} \tag{A.4}$$

and the following re-scaled state variables:

$$\begin{aligned}
\chi_i &= \frac{m_i k_i}{b_i} x_i & i &\in P, O, C,\\
\tau_i &= k_i t & i &\in P, O, C.
\end{aligned} \tag{A.5}$$

Appendix B: Sufficient conditions for the preswitch

Let the preswitch dynamics (2.12) be rewritten in a dimensionless form, for $n = 2$ as

$$\frac{d}{d\tau}x = \frac{\alpha x^2 + 1}{1 + \zeta_D + \gamma x^2} - x, \tag{B.6}$$

by substituting $x = \frac{m_P k_P}{b_P} x_P, \tau = k_P t, \alpha = \frac{a_P b_P}{m_P^2 k_P^2}, \gamma = \frac{c_{PP} b_P^2}{m_P^3 k_P^2}, \zeta_D = \frac{u_D}{b_P}$.

(I) First, we derive necessary and sufficient conditions for (B.6), $u_D = 0$, having exactly two (nonnegative) stable steady states, which is equivalent to the existence of exactly three solutions x to

$$\frac{d}{d\tau}x = \frac{\alpha x^2 + 1}{1 + \gamma x^2} - x \overset{!}{=} 0. \tag{B.7}$$

All solutions satisfy $x \geq 0$ since the parameters $\alpha, \gamma > 0$. This (B.7) is equivalent to

$$p(x) := \gamma x^3 - \alpha x^2 + x - 1 \overset{!}{=} 0, \tag{B.8}$$

which has exactly three solutions if and only if the equation

$$\frac{\partial}{\partial x} p(x) = 3\gamma x^2 - 2\alpha x + 1 = 0 \tag{B.9}$$

(i) has two solutions $x_{1,2} = \frac{\alpha \pm \sqrt{\alpha^2 - 3\gamma}}{3\gamma} \geq 0$ and (ii) $p(x_1) < 0, p(x_2) > 0$. It holds that

(i) $\Leftrightarrow$ $\alpha > \sqrt{3\gamma}$.
(ii) $\Leftrightarrow$ $\left(0 < \alpha \leq \frac{1}{4} \wedge 0 < \gamma < (s_1 + s_2)\right) \vee$ $\left(\frac{1}{4} < \alpha \leq \frac{1}{3} \wedge (s_1 - s_2) < \gamma < (s_1 + s_2)\right)$, $\tag{B.10}$

wherein $s_1 = \frac{1}{27}(-2 + 9\alpha)$ and $s_2 = \frac{2}{27}\sqrt{1 - 9\alpha + 27\alpha^2 - 27\alpha^3}$.

Thus, for parameters α, γ satisfying (B.10), it is guaranteed that x_P can have an "on" and an "off" state.

(II) Second, we proof that there always exists a finite stimulus $\zeta := \zeta_D^{\text{crit}} > 0$ such that there is exactly one stable steady state solution x to

$$\frac{d}{d\tau}x = \frac{\alpha x^2 + 1}{1 + \zeta + \gamma x^2} - x \overset{!}{=} 0. \tag{B.11}$$

This is equivalent to

$$q(x) := \gamma x^3 - \alpha x^2 + (\zeta + 1)x - 1 \overset{!}{=} 0, \tag{B.12}$$

which in turn has exactly one solution if

$$\frac{\partial}{\partial x} q(x) = 3\gamma x^2 - 2\alpha x + (\zeta + 1) = 0 \tag{B.13}$$

has at most one solution.

It can be shown that this holds if and only if $\alpha^2 < 3\gamma(\zeta + 1)$, that is if and only if $\zeta > \frac{\alpha^2}{3\gamma} - 1$. Since from (I) one knows $1 < \frac{\alpha^2}{3\gamma} < \infty$, it is $0 < \zeta$ and there always exists a $\zeta^{\text{crit}} < \infty$ such that there is exactly one stable steady state. Thus, there is always a stimulus ζ_D that can irreversibly switch off x_P.

Appendix C: Proof that the truncation error of the truncated DsC model and of the truncated DsCL model is upper bounded

We prove that the truncation error (4.23) for truncating the ODE system (4.13) with $i \in \{0, \ldots, S-1\}^K$, is upper bounded.

For this purpose, we write a system for the dynamics of $\underline{N}(i|t) := \sum_{j=1}^{J} N(i, j|t)$,

$t \in (T_{k-1}, T_k]$:

$$\frac{d\underline{N}(i|t)}{dt} = -\underline{\beta}_i(t)\underline{N}(i|t) - \underline{\alpha}_i(t)\underline{N}(i|t) + \begin{cases} 0 & , i_k = 0 \\ 2\underline{\alpha}_{i-e_k^T}(t)\underline{N}(i - e_k^T|t) & , i_k \geq 1 \end{cases} \tag{C.14}$$

with initial conditions

$$\mathbf{i} = \mathbf{0}^T : \underline{N}(\mathbf{0}^T|0) = \underline{N}_{\text{ini}} := \sum_{j=1}^{J} N_{\text{ini},j}, \ \forall \mathbf{i} \neq \mathbf{0}^T : \underline{N}(\mathbf{i}|0) = 0, \tag{C.15}$$

and the parameters, relating to the original system (4.13),

$$\underline{\beta}_{\mathbf{i}}(t) := \frac{\sum_{j_1=1}^{J} \beta_{\mathbf{i}}^{j_1}(t) N(\mathbf{i}, j_1|t)}{\sum_{j_1=1}^{J} N(\mathbf{i}, j_1|t)} \quad , \quad \underline{\alpha}_{\mathbf{i}}(t) := \frac{\sum_{j_1=1}^{J} \alpha_{\mathbf{i}}^{j_1}(t) N(\mathbf{i}, j_1|t)}{\sum_{j_1=1}^{J} N(\mathbf{i}, j_1|t)}. \tag{C.16}$$

Furthermore, we define a bounding system in the style of Hasenauer et al. (2012a) but extended to K time intervals,

$$t \in (T_{k-1}, T_k] :$$
$$\frac{dB(\mathbf{i}|t)}{dt} = -\beta_{\text{inf}} B(\mathbf{i}|t) - \alpha_{\text{inf}}^{\text{eff}} B(\mathbf{i}|t) + \begin{cases} 0 & , \ i_k = 0 \\ 2\alpha_{\text{sup}}^{\text{eff}} B(\mathbf{i} - \mathbf{e}_k^T|t) & , \ i_k \geq 1 \end{cases}, \tag{C.17}$$

with initial conditions

$$\mathbf{i} = \mathbf{0}^T : B(\mathbf{0}^T|0) = \underline{N}_{\text{ini}}, \ \forall \mathbf{i} \neq \mathbf{0}^T : B(\mathbf{i}|0) = 0, \tag{C.18}$$

and with parameters

$$\beta_{\text{inf}} := \inf_{\substack{\tau \in (0,T_K] \\ \mathbf{i} \in \mathbb{N}_0^K \\ j \in \{1,\dots,J\}}} \left\{ \beta_{\mathbf{i}}^{j}(\tau) \right\}$$

$$\alpha_{\text{inf}}^{\text{eff}} := \inf_{\substack{\tau \in (0,T_K] \\ \mathbf{i} \in \mathbb{N}_0^K \\ j \in \{1,\dots,J\}}} \left\{ \alpha_{\mathbf{i}}^{j}(\tau) \right\} \tag{C.19}$$

$$\alpha_{\text{sup}}^{\text{eff}} := \sup_{\substack{\tau \in (0,T_K] \\ \mathbf{i} \in \mathbb{N}_0^K \\ j \in \{1,\dots,J\}}} \left\{ \alpha_{\mathbf{i}}^{j}(\tau) \right\}.$$

These parameters are related to the parameters (C.16) of the system (C.14) for all $\mathbf{i}, \tau$ via:

$$
\inf_{j \in \{1,...,J\}} \{\beta_{\mathbf{i}}^j(\tau)\} = \frac{\inf_{j \in \{1,...,J\}} \{\beta_{\mathbf{i}}^j(\tau)\} \sum_{j_1=1}^{J} N(\mathbf{i}, j_1 | \tau)}{\sum_{j_1=1}^{J} N(\mathbf{i}, j_1 | \tau)} \leq \frac{\sum_{j_1=1}^{J} \beta_{\mathbf{i}}^{j_1}(\tau) N(\mathbf{i}, j_1 | \tau)}{\sum_{j_1=1}^{J} N(\mathbf{i}, j_1 | \tau)} = \underline{\beta}_{\mathbf{i}}(\tau)
$$

$$
\inf_{j \in \{1,...,J\}} \{\alpha_{\mathbf{i}}^j(\tau)\} = \frac{\inf_{j \in \{1,...,J\}} \{\alpha_{\mathbf{i}}^j(\tau)\} \sum_{j_1=1}^{J} N(\mathbf{i}, j_1 | \tau)}{\sum_{j_1=1}^{J} N(\mathbf{i}, j_1 | \tau)} \leq \frac{\sum_{j_1=1}^{J} \alpha_{\mathbf{i}}^{j_1}(\tau) N(\mathbf{i}, j_1 | \tau)}{\sum_{j_1=1}^{J} N(\mathbf{i}, j_1 | \tau)} = \underline{\alpha}_{\mathbf{i}}(\tau)
$$

$$
\sup_{j \in \{1,...,J\}} \{\alpha_{\mathbf{i}}^j(\tau)\} = \frac{\sup_{j \in \{1,...,J\}} \{\alpha_{\mathbf{i}}^j(\tau)\} \sum_{j_1=1}^{J} N(\mathbf{i}, j_1 | \tau)}{\sum_{j_1=1}^{J} N(\mathbf{i}, j_1 | \tau)} \geq \frac{\sum_{j_1=1}^{J} \alpha_{\mathbf{i}}^{j_1}(\tau) N(\mathbf{i}, j_1 | \tau)}{\sum_{j_1=1}^{J} N(\mathbf{i}, j_1 | \tau)} = \overline{\alpha}_{\mathbf{i}}(\tau).
$$

$$\text{(C.20)}$$

With this, for all $\mathbf{i}, t$ the outfluxes from $B(\mathbf{i}|t)$ are always less or equal than the outfluxes from $\underline{N}(\mathbf{i}|t)$, and the influxes into $B(\mathbf{i}|t)$ are always greater or equal than the influxes into $\underline{N}(\mathbf{i}|t)$. Therefore, it holds by a similar argument as in Hasenauer et al. (2012a) that

$$
\forall \mathbf{i}, t \geq 0 : \quad \underline{N}(\mathbf{i}|t) \leq B(\mathbf{i}|t). \tag{C.21}
$$

Thus, it also holds that the bounding system gives an upper bound for the sums $\forall k, t \in (T_{k-1}, T_k]$:

$$
\sum_{i_1=S}^{\infty} \cdots \sum_{i_k=S}^{\infty} \underline{N}(\mathbf{i}|t) \leq \sum_{i_1=S}^{\infty} \cdots \sum_{i_k=S}^{\infty} B(\mathbf{i}|t)
$$

$$
M(t) = \sum_{i_1=0}^{\infty} \cdots \sum_{i_k=0}^{\infty} \underline{N}(\mathbf{i}|t) \leq B(t) = \sum_{i_1=0}^{\infty} \cdots \sum_{i_k=0}^{\infty} B(\mathbf{i}|t). \tag{C.22}
$$

Let the truncated overall number of cells be denoted by

$$
\hat{M}_S(t) := \sum_{i_1=0}^{S-1} \cdots \sum_{i_k=0}^{S-1} \underline{N}(\mathbf{i}|t). \tag{C.23}
$$

Because $\sum_{i_1=S}^{\infty} \cdots \sum_{i_k=S}^{\infty} \underline{N}(\mathbf{i}|t) = M(t) - \hat{M}_S(t)$, which is the numerator of the truncation error (4.23), this gives the upper bound for the truncation error

$$
\frac{M(t) - \hat{M}_S(t)}{\underline{N}_{\text{ini}}} \leq E_S(t) = \frac{B(t) - \sum_{i_1=0}^{S-1} \cdots \sum_{i_k=0}^{S-1} B(\mathbf{i}|t)}{\underline{N}_{\text{ini}}}. \tag{C.24}
$$

The upper bound can be calculated since the solution of the bounding system (C.17) can be solved, analoguously to the solution (E.36) derived in Appendix E, just with considering $\alpha_{\text{sup}}^{\text{eff}}$ and $\alpha_{\text{inf}}^{\text{eff}}$ separately:

$$\forall t \in (T_{k-1}, T_k]: \ B(\mathbf{i}|t) = \begin{cases} \dfrac{(2\alpha_{\text{sup}}^{\text{eff}})^{\left(\sum_{l=1}^{k} i_l\right)} \prod_{l=1}^{k-1}(T_l-T_{l-1})^{i_l}(t-T_{k-1})^{i_k}}{\prod_{l=1}^{k} i_l!} \\ \qquad\qquad\qquad\qquad\qquad \cdot e^{-(\alpha_{\text{inf}}^{\text{eff}}+\beta_{\text{inf}})t}\underline{N}_{\text{ini}} \quad , \ \text{if } \forall \nu > k : i_\nu = 0 \\ 0 \qquad\qquad\qquad\qquad\qquad\qquad\qquad\qquad\ , \ \text{otherwise} \end{cases} \ ,$$

$$\text{(C.25)}$$

The upper bound (C.24) implies that, if the model can be decomposed as by (4.12), also the truncation error for the overall model solution, $\|m(x|t) - \hat{m}_S(x|t)\|/\|m(x|0)\|$, is upper bounded by the same error bound. With the decomposition property (4.12), it is $n(x, \mathbf{i}, j|t) = N(\mathbf{i}, j|t)p(x|\mathbf{i}, t)$, thus we can rewrite, similarly to Hasenauer et al. (2012a),

$$\|m(x|t) - \hat{m}_S(x|t)\| = \sum_{i_1=S}^{\infty} \cdots \sum_{i_k=S}^{\infty} \underline{N}(\mathbf{i}|t) \int_{\mathbb{R}_+} p(x|\mathbf{i}, t)$$

$$= \sum_{i_1=S}^{\infty} \cdots \sum_{i_k=S}^{\infty} \underline{N}(\mathbf{i}|t)$$

$$= \sum_{i_1=0}^{\infty} \cdots \sum_{i_k=0}^{\infty} \underline{N}(\mathbf{i}|t) - \sum_{i_1=0}^{S-1} \cdots \sum_{i_k=0}^{S-1} \underline{N}(\mathbf{i}|t)$$

$$= M(t) - \hat{M}_S(t). \qquad\qquad \text{(C.26)}$$

From this it follows directly that (C.24) is also an upper bound for the truncation error of $\|m(x|t) - \hat{m}_S(x|t)\|/\|m(x|0)\|$.

For $S \to \infty$, it is $\sum_{i_1=S}^{\infty} \cdots \sum_{i_k=S}^{\infty} \underline{N}(\mathbf{i}|t) \to M(t)$ by definition, thus $M(t) \to \hat{M}_S(t)$, and thus the truncation error goes to zero.

Appendix D: Proof of the solution for a DsC model with division-dependent parameters

Here we check Proposition 4.3: If (4.24) holds (that is no transitions between cell types), and division and death rates are constant within each time interval, $\forall \mathbf{i}, j, t: \ \alpha_{\mathbf{i}}^j(t) = \alpha_{\mathbf{i}}^j, \beta_{\mathbf{i}}^j(t) = \beta_{\mathbf{i}}^j$, and furthermore $\forall j, \mathbf{i}', \mathbf{i}'', \mathbf{i}' \neq \mathbf{i}'' : \alpha_{\mathbf{i}'}^j + \beta_{\mathbf{i}'}^j \neq \alpha_{\mathbf{i}''}^j + \beta_{\mathbf{i}''}^j$, then the solution of (4.13), (4.14) is given by

$$\forall t \in (T_{k-1}, T_k]: \ N(\mathbf{i}, j | t) = \begin{cases} 2^{(\sum_{\nu=1}^{k} i_\nu)} \left(\prod_{\nu=1}^{k} C_{\mathbf{i},j}^{\nu} \right) \left(\prod_{\nu=1}^{k-1} D_{\mathbf{i},j}^{\nu}(T_\nu) \right) D_{\mathbf{i},j}^{k}(t) N_{\text{ini},j} & , \text{ if } \forall \nu > k : i_\nu = \\ 0 & , \text{ otherwise} \end{cases}$$

with $\tilde{\mathbf{i}}^{(\nu,l)} := [i_1, \ldots, i_{\nu-1}, l, 0, \ldots, 0]$,

$$C_{\mathbf{i},j}^{\nu} := \begin{cases} 1 & , \ i_\nu = 0 \\ \prod_{l=1}^{i_\nu} \alpha_{\tilde{\mathbf{i}}^{(\nu,l-1)}}^{j} & , \ i_\nu \geq 1 \end{cases},$$

$$D_{\mathbf{i},j}^{\nu}(t) := \begin{cases} e^{-(\alpha_{\tilde{\mathbf{i}}^{(\nu,0)}}^{j} + \beta_{\tilde{\mathbf{i}}^{(\nu,0)}}^{j})t} & , \ i_\nu = 0 \\ \left(\sum_{l=0}^{i_\nu} \left[\left(\prod_{\substack{j=0 \\ j \neq l}}^{i_\nu} ((\alpha_{\tilde{\mathbf{i}}^{(\nu,j)}}^{j} + \beta_{\tilde{\mathbf{i}}^{(\nu,j)}}^{j}) - (\alpha_{\tilde{\mathbf{i}}^{(\nu,l)}}^{j} + \beta_{\tilde{\mathbf{i}}^{(\nu,l)}}^{j})) \right)^{-1} \right. \\ \left. \qquad\qquad\qquad \cdot e^{-(\alpha_{\tilde{\mathbf{i}}^{(\nu,l)}}^{j} + \beta_{\tilde{\mathbf{i}}^{(\nu,l)}}^{j})t} \right] \right) & , \ i_\nu \geq 1 \end{cases}.$$

(D.27)

For the proof we make use of the existing solution for one time interval, $K = 1$, and one cell type, $J = 1$, which can be found in Hasenauer et al. (2012a). Since in the considered case it is assumed that (4.24) holds, that is there are no transitions between cell types, the solution presented therein can be applied for each individual cell type to obtain the solution for one time interval,

$$i = 0 : N(0, j | t) = e^{-(\beta_0^j + \alpha_0^j)t} N_{\text{ini},j}(0),$$

$$\forall i \geq 1 : N(i, j | t) = \sum_{q=0}^{i} \left(2^{i-q} \left(\prod_{s=q}^{i-1} \alpha_s^j \right) D_{i,q,j}(t) N_{\text{ini},j}(q) \right),$$

(D.28)

$$\text{with } D_{i,q,j}(t) := \sum_{p=q}^{i} \left[\left(\prod_{\substack{r=q \\ r \neq p}}^{i} ((\beta_r^j + \alpha_r^j) - (\beta_p^j + \alpha_p^j)) \right)^{-1} e^{-(\beta_p^j + \alpha_p^j)t} \right],$$

Using this, we now apply an induction proof to extend the result to $K \geq 1$ time intervals. In the base step it is easy to verify that (D.27) holds for the first time interval $k = 1$, for which by (D.28) it is

$$\forall t \in (0, T_1]: \ N(\mathbf{i}, j | t) = \begin{cases} 0 & , \ \exists \nu > 1 : i_\nu \neq 0 \\ e^{-(\alpha_{\mathbf{i}(1,0)}^{j} + \beta_{\mathbf{i}(1,0)}^{j})t} N_{\text{ini},j} & , \ i_1 = 0 \\ 2^{i_1} \left(\prod_{l=1}^{i_1} \alpha_{\mathbf{i}(1,l-1)}^{j} \right) D_{\mathbf{i},j}^{1}(t) N_{\text{ini},j} & , \ i_1 \geq 1 \end{cases}$$

$$= \begin{cases} 2^{i_1} C_{\mathbf{i},j}^{1} D_{\mathbf{i},j}^{1}(t) N_{\text{ini},j} & , \text{ if } \forall \nu > 1 : i_\nu = 0 \\ 0 & , \text{ otherwise} \end{cases}$$

(D.29)

Now the induction step is done from $k = m - 1$ to $k = m$. Therefore, the initial condition for each k-th time interval are

$$N(\tilde{\mathbf{i}}^{(k-1,0)}, j | T_{k-1}).$$

(D.30)

Then, using again the solution for one time interval (D.29) and substituting the appropriate initial conditions, the solution for the m-th time interval can be rewritten

$\forall t \in (T_{m-1}, T_m]$:

$$
N(\mathbf{i}, j|t) = \begin{cases} 0 & , \exists \nu > m : i_\nu \neq 0 \\ e^{-(\alpha^j_{\mathbf{i}(m_-,0)} + \beta^j_{\mathbf{i}(m,0)})t} N(\tilde{\mathbf{i}}^{(m-1,0)}, j|T_{m-1}) & , i_m = 0 \wedge \forall \nu > m : i_\nu = 0, \\ 2^{i_m} \left(\prod_{l=1}^{i_m} \alpha^j_{\mathbf{i}(m,l-1)} \right) D^m_{\mathbf{i},j}(t) & \\ \qquad \cdot N(\tilde{\mathbf{i}}^{(m-1,0)}, j|T_{m-1}) & , i_m \geq 1 \wedge \forall \nu > m : i_\nu = 0, \end{cases}
$$

$$
= \begin{cases} 2^{i_m} C^m_{\mathbf{i},j} D^m_{\mathbf{i},j}(t) N(\tilde{\mathbf{i}}^{(m-1,0)}, j|T_{m-1}) & , \text{if } \forall \nu > m : i_\nu = 0 \\ 0 & , \text{ otherwise} \end{cases} .
$$

(D.31)

This, together with (D.29) and (D.30), and given that (D.27) holds for $k = m - 1$, leads to

$\forall t \in (T_{m-1}, T_m]$:

$$
N(\mathbf{i}, j|t) = \begin{cases} 2^{i_m} C^m_{\mathbf{i},j} D^m_{\mathbf{i},j}(t) \cdot 2^{\sum_{\nu=1}^{m-1} i_\nu} \left(\prod_{\nu=1}^{m-1} C^\nu_{\mathbf{i},j} \right) & \\ \qquad \cdot \left(\prod_{\nu=1}^{m-2} D^\nu_{\mathbf{i},j}(T_\nu) \right) D^{m-1}_{\mathbf{i},j}(T_{m-1}) N_{\text{ini},j} & , \text{if } \forall \nu > m : i_\nu = 0 \\ 0 & , \text{ otherwise} \end{cases}
$$

$$
= \begin{cases} 2^{\sum_{\nu=1}^{m} i_\nu} \left(\prod_{\nu=1}^{m} C^\nu_{\mathbf{i},j} \right) \left(\prod_{\nu=1}^{m-1} D^\nu_{\mathbf{i},j}(T_\nu) \right) D^m_{\mathbf{i},j}(t) N_{\text{ini},j} & , \text{if } \forall \nu > m : i_\nu = 0 \\ 0 & , \text{ otherwise} \end{cases}
$$

(D.32)

which is the solution stated in (D.27) and in Proposition 4.3.

Appendix E: Proof of the solution for a DsC model with division-independent parameters

The statement of Proposition 4.4 is already known for the special cases of one time interval, $K = 1$, as stated by De Boer et al. (2006), Hasenauer et al. (2012a), Revy et al. (2001), and Schittler et al. (2011), and for two time intervals, $K = 2$, as by Ganusov and De Boer (2013). Here we derive the solution for the general case $K \geq 1$ by an induction proof.

In the base step it is easy to verify that (4.26) holds for the first time interval, $k = 1$, for which the solution is

$$
N(i_1, j|t) = \frac{(2\alpha^j t)^{i_1}}{i_1!} e^{-(\alpha^j + \beta^j)t} N_{\text{ini},j}.
$$

(E.33)

Although this solution (E.33) was already presented in Revy et al. (2001), the proof was first shown in Hasenauer et al. (2012a), Lemma 1, Appendix B therein. The base step is the equation for the first time interval, for which from (E.33) it is

$$
\forall t \in (0, T_1] : \; N(\mathbf{i}, j|t) = \begin{cases} 0 & , \exists \nu > 1 : i_\nu \neq 0 \\ e^{-(\alpha^j + \beta^j)t} N_{\text{ini},j}, & , i_1 = 0 \wedge \forall \nu > 1 : i_\nu = 0, \\ \frac{(2\alpha^j t)^{i_1}}{i_1!} e^{-(\alpha^j + \beta^j)t} N_{\text{ini},j} & , i_1 \geq 1 \wedge \forall \nu > 1 : i_\nu = 0, \end{cases}
$$

$$
= \begin{cases} \frac{(2\alpha^j)^{i_1}(t-0)^{i_1}}{i_1!} e^{-(\alpha^j + \beta^j)t} N_{\text{ini},j} & , \forall \nu > k : i_\nu = 0, \\ 0 & , \text{ otherwise} \end{cases}
$$

(E.34)

Now, the induction step is done from $k = m-1$ to $k = m$, for which again the initial conditions are denoted by (D.30). Using the solution for one time interval (E.33) and substituting the appropriate initial conditions, the solution for the m-th time interval can be rewritten

$\forall t \in (T_{m-1}, T_m]$:

$$N(\mathbf{i}, j|t) = \begin{cases} 0 & , \exists \nu > 1 : i_\nu \neq 0 \\ e^{(-(\alpha^j+\beta^j)(t-T_{m-1}))} N(\mathbf{i}, j|T_{m-1}) & , i_m = 0 \wedge \forall \nu > m : i_\nu = 0, \\ \frac{(2\alpha(t-T_{m-1}))^{i_m}}{i_m!} e^{(-(\alpha+\beta)(t-T_{m-1}))} N(\mathbf{i}, j|T_{m-1}) & , i_m \geq 1 \wedge \forall \nu > m : i_\nu = 0, \end{cases}$$

$$= \begin{cases} \frac{(2\alpha(t-T_{m-1}))^{i_{r_1}}}{i_m!} e^{(-(\alpha+\beta)(t-T_{m-1}))} N(\mathbf{i}, j|T_{m-1}) & , \forall \nu > 1 : i_\nu = 0 \\ 0 & , \text{otherwise} \end{cases}.$$

$$(E.35)$$

This, together with (E.34) and (D.30) and given that (E.33) holds for $k = m - 1$, leads to

$\forall t \in (T_{m-1}, T_m]$:

$$N(\mathbf{i}, j|t) = \begin{cases} \frac{(2\alpha^j(t-T_{m-1}))^{i_m}}{i_m!} e^{(-(\alpha^j+\beta^j)(t-T_{m-1}))} \\ \quad \cdot \frac{(2\alpha^j)^{\left(\sum_{l=1}^{m-1} i_l\right)} \prod_{l=1}^{m-2}(T_l-T_{l-1})^{i_l}(T_{m-1}-T_{m-2})^{i_{m-1}}}{\prod_{l=1}^{m-1} i_l!} \\ \quad \cdot e^{-(\alpha^j+\beta^j)T_{m-1}} N_{\text{ini},j} & , \forall \nu > 1 : i_\nu = 0 \\ 0 & , \text{otherwise} \end{cases}$$

$$= \begin{cases} \frac{(2\alpha^j)^{\left(\sum_{l=1}^{m} i_l\right)} \prod_{l=1}^{m-1}(T_l-T_{l-1})^{i_l}(t-T_{m-1})^{i_m}}{\prod_{l=1}^{m} i_l!} e^{-(\alpha^j+\beta^j)t} N_{\text{ini},j} & , \text{if } \forall \nu > 1 : i_\nu = 0 \\ 0 & , \text{otherwise} \end{cases}$$

$$(E.36)$$

With that, (4.26) is proven.

Appendix F: Derivation of the PBE with stochastic gene expression dynamics

Here we show, in the style of Ramkrishna (2000), how to arrive from (4.35) at the PBE (4.36). The second term on the right hand side of (4.35) can be rewritten, with Ito's formula (Gardiner, 2004),

$$\int_{\mathbb{R}_+^n} \mathbb{E}\left(\frac{dz}{dt}\right) n(z|t)dz = \int_{\mathbb{R}_+^n} \mathbb{E}\left(\nabla\Big(zf(z,u)\Big) + \nabla\Big(\frac{dW_t}{dt}\sigma g(z,u)\Big)\right) + \nabla^2\Big(z\frac{\sigma^2}{2}g(z,u)\Big)\right) n(z|t)dz.$$

$$(F.37)$$

Because W_t is normally-distributed (2.18), it is $\mathbb{E}(W_t) = 0$, and thus the expression simplifies to

$$\int_{\mathbb{R}_+^n} \left(\nabla\Big(zf(z,u)\Big) + \nabla\Big(z\mathbb{E}\Big(\frac{dW_t}{dt}\Big)\Big)\sigma g(z,u) + \nabla^2\Big(z\frac{\sigma^2}{2}g(z,u)\Big)\right) n(z|t)dz$$

$$= \int_{\mathbb{R}_+^n} \left(\nabla\Big(zf(z,u)\Big) + \nabla^2\Big(z\frac{\sigma^2}{2}g(z,u)\Big)\right) n(z|t)dz,$$

$$(F.38)$$

which by partial integration turns into

$$\int_{\mathbb{R}^n_+} z \left(-\nabla \left(f(z,u)n(z|t) \right) + \frac{\sigma^2}{2} \nabla^2 \left(g(z,u)n(z|t) \right) \right) dz. \tag{F.39}$$

Inserting these results for the right hand term in (4.35), and taking all to the left hand side, one arrives at

$$\int_{\mathbb{R}^n_+} z \left(\frac{\partial}{\partial t} n(z|t) - \left(\text{birth}(n(z|t)) - \text{loss}(n(z|t)) \right) + \nabla \left(f(z,u)n(z|t) \right) - \frac{\sigma^2}{2} \nabla^2 \left(g(z,u)n(z|t) \right) \right) dz = 0 \tag{F.40}$$

This must hold in general independently of the value of z, so we know that the expression within the brackets must be zero, thus

$$\frac{\partial}{\partial t} n(z|t) = -\nabla \left(f(z,u)n(z|t) \right) + \frac{\sigma^2}{2} \nabla^2 \left(g(z,u)n(z|t) \right) + \left(\text{birth}(n(z|t)) - \text{loss}(n(z|t)) \right). \tag{F.41}$$

With this we have derived the gene expression dynamics part of the population balance equation.

Appendix G: Matching of fluxes between the C model and the gene switch structured population model

Matching the fluxes of cell type transitions to the respective fluxes in the C model,

$$\bar{\delta}^{j_2 j_1}(t)\overline{N}(j_1|t) = \max \left\{ 0, \int_{\Omega_{j_1}} \left(-\left(f(z,u)n(z|t) \right) + \frac{\sigma^2}{2} \frac{\partial}{\partial z} \left(g(z,u)n(z|t) \right) \right) dz \right\}, \tag{G.42}$$

yields for the cell type transition fluxes from (4.43) and (4.45), as well as from (4.44) and (4.46),

$$\bar{\delta}^{12}(t)N_2(t) = \max \left\{ 0, -f(z^c,u)n(z^c|t) + \frac{\sigma^2}{2} \frac{\partial(g(z,u)n(z|t))}{\partial z} \Big|_{z^c} \right\}$$
$$\bar{\delta}^{21}(t)N_1(t) = \max \left\{ 0, f(z^c,u)n(z^c|t) - \frac{\sigma^2}{2} \frac{\partial(g(z,u)n(z|t))}{\partial z} \Big|_{z^c} \right\}, \tag{G.43}$$

wherein the regularity conditions (4.40) were included.

Let us use $\Delta_f(z,u) := f(z,u) - f(z,0)$ which was defined in order to separate the dynamics induced by a nonzero input. Then, the terms in (G.43) can be rewritten and simplified as follows:

$$f(z^c,u)n(z^c|t) = \underbrace{f(z^c,0)}_{=0} n(z^c|t) + \Delta_f(z^c,u)n(z^c|t) = \Delta_f(z^c,u)n(z^c|t) \tag{G.44}$$

which, in case of no input $u = 0$, equals 0.

Taking this together, (G.43) becomes

$$\bar{\delta}^{12}(t)\overline{N}(2|t) = \max \left\{ 0, -\Delta_f(z^c,u)n(z^c|t) + \frac{\sigma^2}{2} \frac{\partial(g(z,u)n(z|t))}{\partial z} \Big|_{z^c} \right\}$$
$$\bar{\delta}^{21}(t)\overline{N}(1|t) = \max \left\{ 0, \Delta_f(z^c,u)n(z^c|t) - \frac{\sigma^2}{2} \frac{\partial(g(z,u)n(z|t))}{\partial z} \Big|_{z^c} \right\}. \tag{G.45}$$

Appendix H: Solution of (5.21) with division-dependent parameters

The more general case with division-dependent parameters is given by (5.21). Here we show that if

$$\forall i_1 \neq i_2 : (\alpha_{i_1}^1 + \beta_{i_1}^1) \neq (\alpha_{i_2}^1 + \beta_{i_2}^1) \text{ and } \beta_{i_1}^2 \neq (\alpha_{i_2}^1 + \beta_{i_2}^1), \tag{H.46}$$

then the solution of (5.21) is

$$N(i,1|t) = 2^i \left(\prod_{l=0}^{i-1} \alpha_l^1 w_l^{11} \right) \sum_{j=0}^{i} \left(\frac{e^{-(\beta_j^1 + \alpha_j^1)t}}{\prod_{\substack{k=0 \\ k \neq j}}^{i} \left((\beta_k^1 + \alpha_k^1) - (\beta_j^1 + \alpha_j^1) \right)} \right) N_{\text{ini}},$$

$$N(i,2|t) = 2^i \alpha_{i-1}^1 w_{i-1}^{21} \left(\prod_{l=0}^{i-2} \alpha_l^1 w_l^{11} \right) \left(\sum_{j=0}^{i-1} \left(\frac{e^{-(\beta_j^1 + \alpha_j^1)t}}{\prod_{\substack{k=0 \\ k \neq j}}^{i-1} \left((\beta_k^1 + \alpha_k^1) - (\beta_j^1 + \alpha_j^1) \right) \left(\beta_i^2 - (\beta_j^1 + \alpha_j^1) \right)} \right) \right.$$

$$\left. + \frac{e^{-\beta_j^2 t}}{\prod_{k=0}^{i-1} \left((\beta_k^1 + \alpha_k^1) - \beta_i^2 \right)} \right) N_{\text{ini}}. \tag{H.47}$$

The Laplace transform of (H.47), which can be achieved via partial fraction decomposition and induction, is

$$\mathcal{N}(i,1|s) = 2^i \left(\prod_{l=0}^{i-1} \alpha_l^1 w_l^{11} \right) \frac{1}{\prod_{j=0}^{i} (s + \beta_j^1 + \alpha_j^1)} N_{\text{ini}},$$

$$\mathcal{N}(i,2|s) = \frac{2\alpha_{i-1}^1 w_{i-1}^{11}}{(s + \beta_i^2)} \underbrace{2^{i-1} \left(\prod_{l=0}^{i-2} \alpha_l^1 w_l^{11} \right) \frac{1}{\prod_{j=0}^{i-1} (s + \beta_j^1 + \alpha_j^1)} N_{\text{ini}}}_{= \mathcal{N}(i-1,1|s)}. \tag{H.48}$$

• **Base step:** For $i = 0,1$ and $i = 1,2$ it can be shown easily by solving (5.21) in the frequency domain that $\mathcal{N}(0,1|s)$, $\mathcal{N}(1,1|s)$ and $\mathcal{N}(1,2|s)$, $\mathcal{N}(2,2|s)$ are according to (H.48).

• **Induction step:** Given that (H.48) holds for $i = k-1$, it is easy to show that (H.48) holds for $i = k$ by induction:

$$\mathcal{N}(k,1|s) = \frac{2\alpha_{k-1}^1 w_{k-1}^{11}}{(s + \beta_k^1 + \alpha_k^1)} \mathcal{N}(k-1,1|s)$$

$$= \frac{2\alpha_{k-1}^1 w_{k-1}^{11}}{(s + \beta_k^1 + \alpha_k^1)} 2^{k-1} \left(\prod_{l=0}^{k-2} \alpha_l^1 w_l^{11} \right) \frac{1}{\prod_{j=0}^{k-1} (s + \beta_j^1 + \alpha_j^1)} N_{\text{ini}}$$

$$= 2^k \left(\prod_{l=0}^{k-1} \alpha_l^1 w_l^{11} \right) \frac{1}{\prod_{j=0}^{k} (s + \beta_j^1 + \alpha_j^1)} N_{\text{ini}}. \tag{H.49}$$

$$\mathcal{N}(k,2|s) = \frac{2\alpha_{k-1}^1 w_{k-1}^{11}}{(s + \beta_k^2)} \mathcal{N}(k-1,1|s)$$

$$= \frac{2\alpha_{k-1}^1 w_{k-1}^{11}}{(s + \beta_k^2)} 2^{k-1} \left(\prod_{l=0}^{k-2} \alpha_l^1 w_l^{11} \right) \frac{1}{\prod_{j=0}^{k-1} (s + \beta_j^1 + \alpha_j^1)} N_{\text{ini}}.$$

Transformation back into the time domain yields $N(k,1|t)$, $N(k,2|t)$ as given in (H.47). This concludes the proof.

Appendix I: Solution of (5.21) with division-independent parameters

Here we show that the model (5.21) with division-independent parameters, $\forall i : \beta_i^1 = \beta^1$, $\beta_i^2 = \beta^2$, $\alpha_i^1 = \alpha^1$, $w_i^{11} = w^{11}$, $w_i^{21} = w^{21}$, has the solution (5.23).

The Laplace transform of $N(i, 1|t)$ in the given model equations is

$$\mathcal{N}(i, 1|s) = \frac{(2\alpha^1 w^{11})^i}{(s + \beta^1 + \alpha^1)^{i+1}} N_{\text{ini}}. \tag{I.50}$$

First, we show that the first equation in (5.23) for $N(i, 1|t)$, holds:
- Base step: For $i = 0$ and $i = 1$ it can be shown easily that $\mathcal{N}(0, 1|s)$, $\mathcal{N}(1, 1|s)$ are according to (I.50).
- Induction step: Given that (I.50) holds for $i = k - 1$, it is easy to show that (5.23) holds for $i = k$ by induction:

$$
\begin{aligned}
\mathcal{N}(k, 1|s) &= \frac{2\alpha^1 w^{11}}{(s + \beta^1 + \alpha^1)} \frac{(2\alpha^1 w^{11})^{k-1}}{(s + \beta^1 + \alpha^1)^k} N_{\text{ini}} \\
&= \frac{(2\alpha^1 w^{11})^k}{(s + \beta^1 + \alpha^1)^{k+1}} N_{\text{ini}}.
\end{aligned}
\tag{I.51}
$$

Transformation back into the time domain yields $N(i, 1|t)$ in (5.23).

Second, we consider the second equation in (5.23) for $N(i, 2|t)$. It follows directly from considering $N(i - 1, 1|t)$ as an input to $N(i, 2|t)$ and writing

$$
\begin{aligned}
N(i, 2|t) &= \int_0^t e^{-\beta^2(t-\tau)} 2\alpha^1 w^{11} N(i-1, 1|t) d\tau \\
&= 2\alpha^1 w^{11} e^{-\beta^2 t} \int_0^t e^{\beta^2 \tau} \frac{(2w^{11}\alpha^1 \tau)^{i-1}}{(i-1)!} e^{-(\beta^1+\alpha^1)\tau} d\tau \; N_{\text{ini}} \\
&= \frac{(2w^{11}\alpha^1)^i \frac{w^{21}}{w^{11}}}{(i-1)!} e^{-\beta^2 t} \int_0^t \tau^{i-1} e^{(\beta^2-\beta^1-\alpha^1)\tau} d\tau \; N_{\text{ini}}
\end{aligned}
\tag{I.52}
$$

which is the second equation $N(i, 2|t)$ in (5.23).

Appendix J: Total number of stem cells and committed cells

To derive an expression for the total amount of stem cells, we sum up and employ the exponential power series:

$$\bar{N}(1|t) := \sum_{i=0}^{\infty} N(i, 1|t) = \underbrace{\sum_{i=0}^{\infty} \frac{(2w^{11}\alpha^1 t)^i}{i!}}_{=e^{2w^{11}\alpha^1 t}} e^{-(\beta^1+\alpha^1)t} N_{\text{ini}} = e^{((2w^{11}-1)\alpha^1 - \beta^1)t} N_{\text{ini}}. \tag{J.53}$$

To derive the total amount of committed cells, we interchange summation and integration, and employ the exponential power series:

$$
\bar{N}(2|t) := \sum_{i=1}^{\infty} N(i, 2|t) = \sum_{i=1}^{\infty} \frac{(2w^{21}\alpha^1)^i \frac{w^{21}}{w^{11}}}{(i-1)!} e^{-\beta^2 t} \int_0^t \tau^{i-1} e^{(\beta^2 - \beta^1 - \alpha^1)\tau} d\tau \, N_{\text{ini}}
$$

$$
= 2w^{21}\alpha^1 e^{-\beta^2 t} N_{\text{ini}} \int_0^t \sum_{i=1}^{\infty} \frac{(2w^{11}\alpha^1 \tau)^{i-1}}{(i-1)!} e^{(\beta^2 - \beta^1 - \alpha^1)\tau} d\tau
$$

$$
= 2w^{21}\alpha^1 e^{-\beta^2 t} N_{\text{ini}} \int_0^t \underbrace{\sum_{i=0}^{\infty} \frac{(2w^{11}\alpha^1 \tau)^i}{i!}}_{=e^{2w^{11}\alpha^1 \tau}} e^{(\beta^2 - \beta^1 - \alpha^1)\tau} d\tau
$$

$$
\begin{cases}
= \frac{2w^{21}\alpha^1}{\beta^2 + (2w^{11}-1)\alpha^1 - \beta^1} e^{-\beta^2 t} \left(e^{(\beta^2 + (2w^{11}-1)\alpha^1 - \beta^1)t} - 1 \right) N_{\text{ini}} \\
= \frac{2w^{21}\alpha^1}{\beta^2 + (2w^{11}-1)\alpha^1 - \beta^1} \left(e^{((2w^{11}-1)\alpha^1 - \beta^1)t} - e^{-\beta^2 t} \right) N_{\text{ini}} \\
\hspace{6cm} \text{, if } \beta^2 + (2w^{11}-1)\alpha^1 - \beta^1 \neq 0 \\
= 2w^{21}\alpha^1 t e^{-\beta^2 t} N_{\text{ini}} \hspace{3.3cm} \text{, if } \beta^2 + (2w^{11}-1)\alpha^1 - \beta^1 = 0.
\end{cases}
$$

$$(\text{J.54})$$

Appendix K: Committed-to-stem cells ratio and effect of parameters

The growth behavior of the committed-to-stem cells ratio is analyzed as follows, using the results (J.53) and (J.54):

Case (i):

$$
\beta^2 + (2w^{11} - 1)\alpha^1 - \beta^1 > 0
$$

$$
\frac{\bar{N}(2|t)}{\bar{N}(1|t)} = \underbrace{\frac{2w^{21}\alpha^1}{\beta^2 + (2w^{11} - 1)\alpha^1 - \beta^1}}_{>0} \underbrace{\left(1 - e^{-(\beta^2 + (2w^{11}-1)\alpha^1 - \beta^1)t} \right)}_{\to 1 \ (t \to \infty)}
$$

$$
\to \frac{2w^{21}\alpha^1}{\beta^2 + (2w^{11} - 1)\alpha^1 - \beta^1} \quad \text{as } t \to \infty.
$$

$$(\text{K.55})$$

Case (ii):

$$
\beta^2 + (2w^{11} - 1)\alpha^1 - \beta^1 = 0
$$

$$
\frac{\bar{N}(2|t)}{\bar{N}(1|t)} = \frac{2w^{21}\alpha^1 t e^{-\beta^2 t}}{e^{((2w^{11}-1)\alpha^1 - \beta^1)t}}
$$

$$
= 2w^{21}\alpha^1 t e^{-(\beta^2 + (2w^{11}-1)\alpha^1 - \beta^1)t} = 2w^{21}\alpha^1 t.
$$

$$(\text{K.56})$$

Case (iii):

$$\beta^2 + (2w^{11} - 1)\alpha^1 - \beta^1 < 0$$

$$\frac{\bar{N}(2|t)}{\bar{N}(1|t)} = \frac{2w^{21}\alpha^1}{\beta^2 + (2w^{11} - 1)\alpha^1 - \beta^1} \left(1 - e^{-(\beta^2 + (2w^{11} - 1)\alpha^1 - \beta^1)t}\right)$$

$$= \underbrace{\frac{2w^{21}\alpha^1}{\beta^2 + (2w^{11} - 1)\alpha^1 - \beta^1}}_{<0} \underbrace{e^{-(\beta^2 + (2w^{11} - 1)\alpha^1 - \beta^1)t}}_{\geq e^0 = 1} \underbrace{\left(e^{(\beta^2 + (2w^{11} - 1)\alpha^1 - \beta^1)t} - 1\right)}_{\to -1 \ (t \to \infty)}$$

$$\to \underbrace{\frac{-2w^{21}\alpha^1}{\beta^2 + (2w^{11} - 1)\alpha^1 - \beta^1}}_{>0} e^{-(\beta^2 + (2w^{11} - 1)\alpha^1 - \beta^1)t} \quad \text{as } t \to \infty$$

which corresponds to exponential growth.

$$(K.57)$$

The effects of the parameters on the maximum ratio $R_{2/1}^{\max}$ can be studied by the sign of the respective derivative. The particular derivatives are:

$$\frac{\partial R_{2/1}^{\max}}{\partial \alpha^1} = \frac{2w^{21}(\beta^2 - \beta^1)}{(\beta^2 + (2w^{11} - 1)\alpha^1 - \beta^1)^2} \left\{ \begin{array}{ll} > 0 & \text{if } \beta^2 > \beta^1 \\ = 0 & \text{if } \beta^2 = \beta^1 \\ < 0 & \text{if } \beta^2 < \beta^1 \end{array} \right.$$

$$\frac{\partial R_{2/1}^{\max}}{\partial \beta^1} = \frac{2w^{21}\alpha^1}{(\beta^2 + (2w^{11} - 1)\alpha^1 - \beta^1)^2} \quad > 0$$

$$\frac{\partial R_{2/1}^{\max}}{\partial \beta^2} = \frac{-2w^{21}\alpha^1}{(\beta^2 + (2w^{11} - 1)\alpha^1 - \beta^1)^2} \quad < 0$$

$$\frac{\partial R_{2/1}^{\max}}{\partial w^{11}} = \frac{-4w^{21}(\alpha^1)^2}{(\beta^2 + (2w^{11} - 1)\alpha^1 - \beta^1)^2} \quad < 0$$

$$\frac{\partial R_{2/1}^{\max}}{\partial w^{21}} = \frac{2\alpha^1}{\beta^2 + (2w^{11} - 1)\alpha^1 - \beta^1} \quad > 0$$

$$(K.58)$$

For a positive sign, increasing the respective parameter increases the value of $R_{2/1}^{\max}$. For a negative sign, increasing the respective parameter reduces the value of $R_{2/1}^{\max}$.

APPENDIX

Bibliography

B. Alberts, D. Bray, J. Lewis, M. Raff, K. Roberts, and P. Walter. *Molecular biology of the cell*. Garland Science, New York, NY, 4th edition, 2000.

U. Alon. Network motifs: theory and experimental approaches. *Nat. Rev. Genet.*, 8(6):450–461, 2007a.

U. Alon. Simplicity in biology. *Nature*, 446:497, 2007b.

D. Angeli, J.E. Ferrell, and E.D. Sontag. Detection of multistability, bifurcations, and hysteresis in a large class of biological positive-feedback systems. *Proc. Natl. Acad. Sci. U.S.A.*, 101(7):1822–1827, 2004.

A.C. Antoulas. An overview of approximation methods for large-scale dynamical systems. *Annu. Rev. Control*, 29:181–190, 2005a.

A.C. Antoulas. *Approximation of large-scale dynamical systems*. Advances in Design and Control. SIAM, Philadelphia, 2005b.

B. Asquith, C. Debacq, D.C. Macallan, L. Willems, and C.R.M. Bangham. Lymphocyte kinetics: the interpretation of labelling data. *Trends Immunol.*, 23(12):596–601, 2002.

B. Asquith, C. Debacq, A. Florins, N. Gillet, T. Sanchez-Alcaraz, A. Mosley, and L. Willems. Quantifying lymphocyte kinetics in vivo using carboxyfluorescein diacetate succinimidyl ester. *Proc. R. Soc. B: Biological Sciences*, 273(1590):1165–1171, 2006.

R. Assar, A.V. Leisewitz, A. Garcia, N.C. Inestrosa, M.A. Montecino, and D.J. Sherman. Reusing and composing models of cell fate regulation of human bone precursor cells. *Biosystems*, 108(1–3):63–72, 2012.

D. Baksh, L. Song, and R.S. Tuan. Adult mesenchymal stem cells: Characterization, differentiation, and application in cell and gene therapy. *J. Cell. Mol. Med.*, 8(3):301–316, 2004.

G. Balázsi, A. van Oudenaarden, and J.J. Collins. Cellular decision making and biological noise: from microbes to mammals. *Cell*, 144(6):910–925, 2011.

H.T. Banks, K.L Sutton, W.C. Thompson, G. Bocharov, D. Roose, T. Schenkel, and A. Meyerhans. Estimation of cell proliferation dynamics using CFSE data. *Bull. Math. Biol.*, 73(1): 116–150, 2010.

H.T. Banks, K.L. Sutton, W.C. Thompson, G.Bocharov, M. Doumic, T. Schenkel, J. Argilaguet, S. Giest, C. Peligero, and A. Meyerhans. A new model for the estimation of cell proliferation dynamics using CFSE data. *J. Immunol. Meth.*, 373(1-2):143–160, 2011.

H.T. Banks, W.C. Thompson, C. Peligero, S. Giest, J. Argilaguet, and A. Meyerhans. A division-dependent compartmental model for computing cell numbers in CFSE-based lymphocyte proliferation assays. *Math. Biosci. Eng.*, 9(4):699–736, 2012.

H.T. Banks, A. Choi, T. Huffman, J. Nardini, L. Poag, and W.C. Thompson. Quantifying CFSE label decay in flow cytometry data. *Appl. Math. Lett.*, 26(5):571–577, 2013.

M. Bansal, V. Belcastro, A. Ambesi-Impiombato, and D. di Bernardo. How to infer gene networks from expression profiles. *Mol. Syst. Biol.*, 3:78, 2007.

Y. Bar-Yam, D. Harmon, and B. de Bivort. Systems biology. attractors and democratic dynamics. *Science*, 323(5917):1016–1017, 2009.

F.P. Barry and J.M. Murphy. Mesenchymal stem cells: clinical applications and biological characterization. *Int. J. Biochem. Cell Biol.*, 36(4):568–584, 2004.

A. Bertuzzi and A. Gandolfi. A model for estimating cell kinetic parameters of experimental tumours studied by BrdUrd labelling and flow cytometry. *Archives of Control Sciences*, 2000.

J.J. Bird, D.R. Brown, A.C. Mullen, N. H. Moskowitz, M.A. Mahowald, J.R. Sider, T.F. Gajewski, C.R. Wang, and S.L. Reiner. Helper T cell differentiation is controlled by the cell cycle. *Immunity*, 9(2):229–237, 1998.

C. Blanpain and B.D. Simons. Unravelling stem cell dynamics by lineage tracing. *Nat. Rev. Mol. Cell Biol.*, 14(8):489–502, 2013.

S. Bonhoeffer, H. Mohri, D. Ho, and A.S. Perelson. Quantification of cell turnover kinetics using 5-Bromo-2'-deoxyuridine. *J. Immunol.*, 164(10):5049–5054, 2000.

C. Breindl and F. Allgöwer. Verification of multistability in gene regulation networks: A combinatorial approach. In *Proc. 48th IEEE Conf. Decision and Control (CDC)*, pages 5637–5642, Shanghai, China, 2009.

C. Breindl, S. Waldherr, and F. Allgöwer. A robustness measure for the stationary behavior of qualitative gene regulation networks. In *Proc. of the 11th Symposium on Computer Applications in Biotechnology (CAB)*, pages 36–41, Leuven, Belgium, 2010.

C. Breindl, D. Schittler, S. Waldherr, and F. Allgöwer. Structural requirements and discrimination of cell differentiation networks. In *Proc. of the 18th IFAC World Congress*, Milano, Italy, 2011.

P. Buske, J. Galle, N. Barker, G. Aust, H. Clevers, and M. Loeffler. A comprehensive model of the spatio-temporal stem cell and tissue organisation in the intestinal crypt. *PLoS Comput. Biol.*, 7(1):e1001045, 2011.

H.H. Chang, P.Y. Oh, D.E. Ingber, and S.Huang. Multistable and multistep dynamics in neutrophil differentiation. *BMC Cell Biol.*, 7:11, 2006.

H.H. Chang, M. Hemberg, M. Barahona, D.E. Ingber, and S. Huang. Transcriptome-wide noise controls lineage choice in mammalian progenitor cells. *Nature*, 453(7194):544–547, 2008.

R.Z. Chas'minskij. *Stochastic stability of differential equations.* Sijthoff & Noordhoff, 1980.

M. Chaves, E. Farcot, and J.-L. Gouzé. Transition probabilities for piecewise affine models of genetic networks. In *Proc. Int. Symp. Mathematical Theory of Networks and Systems (MTNS 10)*, Budapest, Hungary, 2010.

J.L. Cherry and F.R. Adler. How to make a biological switch. *J. Theor. Biol.*, 203(2):117–133, 2000.

H.-D. Chiang and L. Fekih-Ahmed. Persistence of saddle-node bifurcations for general nonlinear systems under unmodeled dynamics and applications. In *Proc. of 1993 IEEE Intl. Sympos. on Circuits and Systems (ISCAS '93)*, volume 4, pages 2656–2659, 1993.

V. Chickarmane, T. Enver, and C. Peterson. Computational modeling of the hematopoietic erythroid-myeloid switch reveals insights into cooperativity, priming, and irreversibility. *PLoS Comput. Biol.*, 5(1):e1000268, 2009.

O. Cinquin and J. Demongeot. Positive and negative feedback: striking a balance between necessary antagonists. *J. Theor. Biol.*, 216(2):229–241, 2002.

O. Cinquin and J. Demongeot. High-dimensional switches and the modelling of cellular differentiation. *J. Theor. Biol.*, 233(3):391–411, 2005.

M.J. Conboy, A.O. Karasov, and T.A. Rando. High incidence of non-random template strand segregation and asymmetric fate determination in dividing stem cells and their progeny. *PLoS Biol.*, 5(5):e102, 2007.

C. Conradi, D. Flockerzi, J. Raisch, and J. Stelling. Subnetwork analysis reveals dynamic features of complex (bio)chemical networks. *Proc. Natl. Acad. Sci. U.S.A.*, 104(49):19175–19180, 2007.

H. Coskun, T.L.S. Summerfield, D.A. Kniss, and A. Friedman. Mathematical modeling of preadipocyte fate determination. *J. Theor. Biol.*, 265(1):87–94, 2010.

G.J. Darlington, S.E. Ross, and O.A. MacDougald. The role of C/EBP genes in adipocyte differentiation. *J. Biol. Chem.*, 273(46):30057–30060, 1998.

R.J. De Boer and A.J. Noest. T cell renewal rates, telomerase, and telomere length shortening. *J. Immunol.*, 160(12):5832–5837, 1998.

R.J. De Boer and A.S. Perelson. Towards a general function describing T cell proliferation. *J. Theor. Biol.*, 175(4):567–576, 1995.

R.J. De Boer and A.S. Perelson. Quantifying T lymphocyte turnover. *J. Theor. Biol.*, 327:45–87, 2013.

R.J. De Boer, D. Homann, and A.S. Perelson. Different dynamics of CD4+ and CD8+ T cell responses during and after acute lymphocytic choriomeningitis virus infection. *J. Immunol.*, 171(8):3928–3935, 2003a.

R.J. De Boer, H. Mohri, D.D. Ho, and A.S. Perelson. Estimating average cellular turnover from 5-bromo-2'-deoxyuridine (BrdU) measurements. *Proc. R. Soc. Lond. B*, 270:849–858, 2003b.

R.J. De Boer, H. Mohri, D.D. Ho, and A.S. Perelson. Turnover rates of B cells, T cells, and NK cells in simian immunodefficiency virus-infected and uninfected rhesus macaques. *J. Immunol.*, 170:2479–2487, 2003c.

R.J. De Boer, V.V. Ganusov, D. Milutinovic, P.D. Hodgkin, and A.S. Perelson. Estimating lymphocyte division and death rates from CFSE data. *Bull. Math. Biol.*, 68(5):1011–1031, 2006.

R.J. De Boer, A.S. Perelson, and R.M. Ribeiro. Modelling deuterium labelling of lymphocytes with temporal and/or kinetic heterogeneity. *J. R. Soc. Interface*, 9(74):2191–200, 2012.

C. Debacq, B. Asquith, P. Kerkhofs, D. Portetelle, A. Burny, R. Kettmann, and L. Willems. Increased cell proliferation, but not reduced cell death, induces lymphocytosis in bovine leukemia virus-infected sheep. *Proc. Natl. Acad. Sci. U.S.A.*, 99(15):10048–10053, 2002.

C. Debacq, N. Gillet, B. Asquith, M.T. Sanchez-Alcaraz, A. Florins, M. Boxus, I.Schwartz-Cornil, M. Bonneau, G. Jean, P. Kerkhofs, J. Hay, A. Théwis, R. Kettmann, and L. Willems. Peripheral blood B-cell death compensates for excessive proliferation in lymphoid tissues and maintains homeostasis in bovine leukemia virus-infected sheep. *J. Virol.*, 80(19):9710–9719, 2006.

E.K. Deenick, A.V. Gett, and P.D. Hodgkin. Stochastic model of T cell proliferation: A calculus revealing IL-2 regulation of precursor frequencies, cell cycle time, and survival. *J. Immunol.*, 170(10):4963–4972, 2003.

A. Dhooge, W. Govaerts, and Y.A. Kuznetsov. MATCONT: A MATLAB package for numerical bifurcation analysis of ODEs. *ACM Trans. Math. Softw.*, 29(2):141–164, 2003.

H. Drissi, Q. Luc, R. Shakoori, S.C. de Sousa Lopes, J.-Y. Choi, A. Terry, M. Hu, S. Jones, J.C. Neil, J.B. Lian, J.L. Stein, A.J. van Wijnen, and G.S. Stein. Transcriptional autoregulation of the bone related CBFA1/RUNX2 gene. *J. Cell. Physiol.*, 184(3):341–350, 2000.

R. Erban, I.G. Kevrekidis, D. Adalsteinsson, and T.C. Elston. Gene regulatory networks: A coarse-grained, equation-free approach to multiscale computation. *J. Chem. Phys.*, 124(8): 084106, 2006.

L. Fenton. The sum of log-normal probability distributions in scatter transmission systems. *IRE Trans. Commun. Syst.*, 8(1):57–67, 1960.

J.E. Ferrell and E.M. Machleder. The biochemical basis of an all-or-none cell fate switch in Xenopus oocytes. *Science*, 280(5365):895–898, 1998.

J.E. Jr. Ferrell. Bistability, bifurcations, and Waddington's epigenetic landscape. *Curr. Biol.*, 22(11):R458–R466, 2012.

E. Feytmans, D. Noble, and M.C. Peitsch. Genome size and numbers of biological functions. In *Transactions on Computational Systems Biology I*, volume 3380 of *Lecture Notes in Computer Science*, pages 44–49. Springer Berlin Heidelberg, 2005.

A. Florins, N. Gillet, B. Asquith, C. Debacq, G. Jean, I. Schwartz-Cornil, M. Bonneau, A. Burny, M. Reichert, R. Kettmann, and L. Willems. Spleen-dependent turnover of CD11b peripheral blood B lymphocytes in bovine leukemia virus-infected sheep. *J. Virol.*, 80(24): 11998–12008, 2006.

D.V. Foster, J.G. Foster, S. Huang, and S.A. Kauffman. A model of sequential branching in hierarchical cell fate determination. *J. Theor. Biol.*, 260(4):589–597, 2009.

A.G. Fredrickson, D. Ramkrishna, and H.M. Tsuchiya. Statistics and dynamics of procaryotic cell populations. *Math. Biosci.*, 1(3):327–374, 1967.

J. Frith and P. Genever. Transcriptional control of mesenchymal stem cell differentiation. *Transfus. Med. Hemother.*, 35(3):216–227, 2008.

H. Fu, B. Doll, T. McNelis, and J.O. Hollinger. Osteoblast differentiation in vitro and in vivo promoted by osterix. *J. Biomed. Mat. Res. Part A*, 83A(3):770–778, 2007.

V.V. Ganusov and R.J. De Boer. A mechanistic model for BrdU dilution naturally explains labeling data of self-renewing T cell populations. *J. R. Soc. Interface*, 10(78):20120617, 2013.

V.V. Ganusov, S.S. Pilyugin, R.J. De Boer, K. Murali-Krishna, R. Ahmed, and R. Antia. Quantifying cell turnover using CFSE data. *J. Immunol. Meth.*, 298(1-2):183–200, 2005.

C.W. Gardiner. *Handbook of stochastic methods*. Springer Berlin Heidelberg, 3rd edition, 2004.

T.S. Gardner, C.R. Cantor, and J.J. Collins. Construction of a genetic toggle switch in Escherichia coli. *Nature*, 403(6767):339–342, 2000.

A.V. Gett and P.D. Hodgkin. A cellular calculus for signal integration by T cells. *Nat. Immunol.*, 1(3):239–244, 2000.

D. Gillespie. Exact stochastic simulation of coupled chemical reactions. *J. Phys. Chem.*, 81: 2340–2361, 1977.

D. Gillespie. Approximative accelerated stochastic simulation of chemically reacting systems. *J. Chem. Phys.*, 115:1716–1733, 2001.

M. Girgenrath, S. Weng, C.A. Kostek, B. Browning, M. Wang, S.A.N. Brown, J.A. Winkles, J.S. Michaelson, N. Allaire, P. Schneider, M.L. Scott, Y.-M. Hsu, H. Yagita, R.A. Flavell, J.B. Miller, L.C. Burkly, and T.S. Zheng. TWEAK, via its receptor Fn14, is a novel regulator of mesenchymal progenitor cells and skeletal muscle regeneration. *EMBO J.*, 25:5826–5839, 2006.

I. Glauche. *Theoretical studies on the lineage specification of hematopoietic stem cells.* Ph.D. thesis, University of Leipzig, Germany, 2010.

I. Glauche, K. Moore, L. Thielecke, K. Horn, M. Loeffler, and I. Roeder. Stem cell proliferation and quiescence–two sides of the same coin. *PLoS Comput. Biol.*, 5(7):e1000447, 2009.

I. Glauche, M. Herberg, and I. Roeder. Nanog variability and pluripotency regulation of embryonic stem cells–insights from a mathematical model analysis. *PLoS One*, 5(6):e11238, 2010.

A.D. Goldberg, C.D. Allis, and E. Bernstein. Epigenetics: A landscape takes shape. *Cell*, 128: 635–638, 2007.

A. Goldbeter and D.E. Koshland. Ultrasensitivity in biochemical systems controlled by covalent modification. interplay between zero-order and multistep effects. *J. Biol. Chem.*, 259 (23):14441–14447, 1984.

H.G. Gratzner. Monoclonal antibody to 5-bromo- and 5-iododeoxyuridine: A new reagent for detection of DNA replication. *Science*, 218(4571):474–475, 1982.

Z. Grossman, R.B. Herberman, and D.S. Dimitrov. T cell turnover in SIV infection. *Science*, 284(5414):555, 1999.

M. Groszer, R. Erickson, D.D. Scripture-Adams, R. Lesche, A. Trumpp, J.A. Zack, H.I. Kornblum, X. Liu, and H. Wu. Negative regulation of neural stem/progenitor cell proliferation by the Pten tumor suppressor gene in vivo. *Science*, 294(5549):2186–2189, 2001.

M. Gyllenberg. The size and scar distributions of the yeast Saccharomyces cerevisiae. *J. Math. Biol.*, 24:81–101, 1986.

M. Gyllenberg and G.F. Webb. A nonlinear structured population model of tumor growth with quiescence. *J. Math. Biol.*, 28:671–694, 1990.

L.H. Hartwell, J.J. Hopfield, S. Leibler, and A.W. Murray. From molecular to modular cell biology. *Nature*, 402(6761 Suppl):C47–52, 1999.

J. Hasenauer. *Modeling and parameter estimation for heterogeneous cell populations.* Ph.D. thesis, University of Stuttgart, Germany, 2012.

J. Hasenauer, D. Schittler, and F. Allgöwer. Analysis and simulation of division- and label-structured population models. *Bull. Math. Biol.*, 74(11):2692–2732, 2012a.

J. Hasenauer, D. Schittler, and F. Allgöwer. A computational model for proliferation dynamics of division- and label-structured populations. Technical report, arXiv:1202.4923v1 [q-bio.PE], 2012b.

J. Hasty, J. Pradines, M. Dolnik, and J.J. Collins. Noise-based switches and amplifiers for gene expression. *Proc. Natl. Acad. Sci. U.S.A.*, 97(5):2075–2080, 1999.

E.D. Hawkins, M. Hommel, M.L. Turner, F.L. Battye, J.F. Markham, and P.D. Hodgkin. Measuring lymphocyte proliferation, survival and differentiation using CFSE time-series data. *Nat. Protoc.*, 2(9):2057–2067, 2007.

L. Hayflick. The limited in vitro lifetime of human diploid cell strains. *Exp. Cell Res.*, 37(3): 614–636, 1965.

L. Hayflick. Progress in cytogerontology. *Mech. Ageing Dev.*, 9(5–6):393–408, 1979.

T.J. Heino and T.A. Hentunen. Differentiation of osteoblasts and osteocytes from mesenchymal stem cells. *Curr. Stem Cell Res. Ther.*, 3(2):131–145, 2008.

M.K. Hellerstein, R.A. Hoh, M.B. Hanley, D. Cesar, D. Lee, R.A. Neese, and J.M. McCune. Subpopulations of long-lived and short-lived T cells in advanced HIV-1 infection. *J. Clin. Invest.*, 112(6):956–966, 2003.

D.J. Higham. An algorithmic introduction to numerical simulation of stochastic differential equations. *SIAM J. Numer. Anal.*, 43(3):525–546, 2001.

D.J. Higham. Modeling and simulating chemical reactions. *SIAM Rev.*, 50:347–368, 2008.

B.T. Hill, A. Tsuboi, and R. Baserga. Effect of 5-Bromodeoxyuridine on chromatin transcription in confluent fibroblasts. *Proc. Natl. Acad. Sci. U.S.A.*, 71(2):455–459, 1974.

A.L. Hodgkin and A.F. Huxley. A quantitative description of membrane current and its application to conduction and excitation in nerve. *J. Physiol.*, 117(4):500–544, 1952.

T. Holyoake, X. Jiang, C. Eaves, and A. Eaves. Isolation of a highly quiescent subpopulation of primitive leukemic cells in chronic myeloid leukemia. *Blood*, 94(6):2056–2064, 1999.

T. Hong, J. Xing, L. Li, and J. Tyson. A simple theoretical framework for understanding heterogeneous differentiation of CD4+T cells. *BMC Syst. Biol.*, 6(1):66, 2012.

T. Hoshino, T. Nagashima, J. Murovic, E.M. Levin, V.A. Levin, and S.M. Rupp. Cell kinetic studies of in situ human brain tumors with bromodeoxyuridine. *Cytometry*, 6(6):627–632, 1985.

S. Huang. Reprogramming cell fates: reconciling rarity with robustness. *Bioessays*, 31(5): 546–560, 2009.

S. Huang, G. Eichler, Y. Bar-Yam, and D.E. Ingber. Cell fates as high-dimensional attractor states of a complex gene regulatory network. *Phys. Rev. Lett.*, 94(12):128701, 2005.

S. Huang, Y.-P. Guo, G. May, and T. Enver. Bifurcation dynamics in lineage-commitment in bipotent progenitor cells. *Dev. Biol.*, 305(2):695–713, 2007.

O. Ignatyev and V. Mandrekar. Barbashin-Krasovskii theorem for stochastic differential equations. *Proc. Amer. Math. Soc.*, 11:4123–4128, 2010.

L.E. Jones and A.S. Perelson. Opportunistic infection as a cause of transient viremia in chronically infected HIV patients under treatment with HAART. *Bull. Math. Biol.*, 67(6):1227–1251, 2005.

T. Jouini. *Multistabilitätsanalyse von Genregulationssystemen anhand von Konzepten der Mehrgrößenregelung.* Bachelor's thesis, Institute for Systems Theory and Automatic Control (IST), University of Stuttgart, Germany, 2013.

S.A. Kauffman. Metabolic stability and epigenesis in randomly constructed genetic nets. *J. Theor. Biol.*, 22(3):437–467, 1969.

J. Kawaguchi, P.J. Mee, and A.G. Smith. Osteogenic and chondrogenic differentiation of embryonic stem cells in response to specific growth factors. *Bone*, 36(5):758–769, 2005.

N. Kee, S. Sivalingam, R. Boonstra, and J.M. Wojtowicz. The utility of Ki-67 and BrdU as proliferative markers of adult neurogenesis. *J. Neurosci. Meth.*, 115(1):97–105, 2002.

M.J. Kiel, S. He, R. Ashkenazi, S.N. Gentry, M. Teta, J.A. Kushner, T.L. Jackson, and S.J. Morrison. Haematopoietic stem cells do not asymmetrically segregate chromosomes or retain BrdU. *Nature*, 449:238–242, 2007.

D.L. Kline and E.E. Cliffton. The life span of leucocytes in the human. *Science*, 115(2975): 9–11, 1952.

E. Klipp, R. Herwig, A. Kowald, C. Wierling, and H. Lehrach. *Systems biology in practice.* Wiley-VCH Weinheim, 2000.

P.E. Kloeden and E. Platen. *Numerical solution of stochastic differential equations.* Springer Berlin / Heidelberg, 3rd edition, 1999.

J.A. Knoblich. Mechanisms of asymmetric stem cell division. *Cell*, 132(4):583–597, 2008.

J. Krumsiek, S. Pölsterl, D.M. Wittmann, and F.J. Theis. Odefy–from discrete to continuous models. *BMC Bioinf.*, 11:233, 2010.

J. Krumsiek, C. Marr, T. Schroeder, and F.J. Theis. Hierachical differentiation of myeloid progenitors is encoded in the transcription factor network. *PLoS ONE*, 6(8):e22649, 2011.

A. Kuchina, L. Espinar, T. Çagatay, A. Balbin, F. Zhang, A. Alvarado, J. Garcia-Ojalvo, and G.M. Süel. Temporal competition between differentiation programs determines cell fate choice. *Mol. Syst. Biol.*, 7:557, 2011.

D. Kumar and A.B. Lassar. The transcriptional activity of Sox9 in chondrocytes is regulated by RhoA signaling and actin polimerization. *Mol. Cell. Biol.*, 29(15):4262–4273, 2009.

P. Laslo, C.J. Spooner, A. Warmflash, D.W. Lancki, H.-J. Lee, R. Sciammas, B.N. Gantner, A.R. Dinner, and H. Singh. Multilineage transcriptional priming and determination of alternate hematopoietic cell fates. *Cell*, 126(4):755–766, 2006.

H.-Y. Lee and A.S. Perelson. Modeling T cell proliferation and death in vitro based on labeling data: generalizations of the Smith-Martin cell cycle model. *Bull. Math. Biol.*, 70(1):21–44, 2008.

B. Lehner, B. Sandner, J. Marschallinger, C. Lehner, T. Furtner, S. Couillard-Despres, F.J. Rivera, G. Brockhoff, H.-C. Bauer, N. Weidner, and L. Aigner. The dark side of BrdU in neural stem cell biology: detrimental effects on cell cycle, differentiation and survival. *Cell and Tissue Research*, 345(3):313–328, 2011.

K. León, J. Faro, and J. Carneiro. A general mathematical framework to model generation structure in a population of asynchronously dividing cells. *J. Theor. Biol.*, 229(4):455–476, 2004.

C. Li, L. Chen, and K. Aihara. Stability of genetic networks with SUM regulatory logic: Lur'e system and LMI approach. *Circuits and Systems I: Regular Papers, IEEE Transactions on*, 53(11):2451–2458, 2006.

T. Luzyanina, S. Mrusek, J.T. Edwards, D. Roose, S. Ehl, and G. Bocharov. Computational analysis of CFSE proliferation assay. *J. Math. Biol.*, 54(1):57–89, 2007a.

T. Luzyanina, D. Roose, T. Schenkel, M. Sester, S. Ehl, A. Meyerhans, and G. Bocharov. Numerical modelling of label-structured cell population growth using CFSE distribution data. *Theor. Biol. Med. Model.*, 4:26, 2007b.

T. Luzyanina, D. Roose, and G. Bocharov. Distributed parameter identification for a label-structured cell population dynamics model using CFSE histogram time-series data. *J. Math. Biol.*, 59(5):581–603, 2009.

A.B. Lyons. Analysing cell division in vivo and in vitro using flow cytometric measurement of CFSE dye dilution. *J. Immunol. Methods*, 243(1–2):147–154, 2000.

A.B. Lyons and C.R. Parish. Determination of lymphocyte division by flow cytometry. *J. Immunol. Methods*, 171(1):131–137, 1994.

W. Ma, A. Trusina, H. El-Samad, W.A. Lim, and C. Tang. Defining network topologies that can achieve biochemical adaptation. *Cell*, 138(4):760–773, 2009.

B.D. MacArthur, C.P. Please, and R.O.C. Oreffo. Stochasticity and the molecular mechanisms of induced pluripotency. *PLoS One*, 3(8):e3086, 2008.

B.D. MacArthur, A. Ma'ayan, and I.R. Lemischka. Systems biology of stem cell fate and cellular reprogramming. *Nat. Rev. Mol. Cell Biol.*, 10(10):672–681, 2009.

N.V. Mantzaris. Stochastic and deterministic simulations of heterogeneous cell population dynamics. *J. Theor. Biol.*, 241:690–706, 2006.

N.V. Mantzaris. From single-cell genetic architecture to cell population dynamics: Quantitatively decomposing the effects of different population heterogeneity sources for a genetic network with positive feedback architecture. *Biophys. J.*, 92(12):4271–4288, 2007.

A. Marciniak-Czochra, T. Stiehl, A.D. Ho, W. Jaeger, and W. Wagner. Modeling of asymmetric cell division in hematopoietic stem cells–regulation of self-renewal is essential for efficient repopulation. *Stem Cells Dev.*, 18(3):377–385, 2009.

M.R. Martinez, A. Corradin, U. Klein, M.J. Alvarez, G.M. Toffolo, B. di Camillo, A. Califano, and G.A. Stolovitzky. Quantitative modeling of the terminal differentiation of B cells and mechanisms of lymphomagenesis. *Proc. Natl. Acad. Sci. U.S.A.*, 2012.

G. Matera, M. Lupi, and P. Ubezio. Heterogeneous cell response to topotecan in a CFSE-based proliferation test. *Cytometry*, 62(2):118–128, 2004.

P. Metzger, J. Hasenauer, and F. Allgöwer. Modeling and analysis of division-, age-, and label-structured cell populations. In *Proc. 9th Int. Workshop on Computational Systems Biology (WCSB)*, TICSP series # 61, pages 60–63, Ulm, Germany, 2012.

H. Miao, X. Jin, A.S. Perelson, and H. Wu. Evaluation of multitype mathematical models for CFSE-labeling experiment data. *Bull. Math. Biol.*, 74(2):300–326, 2012.

H. Mohri, S. Bonhoeffer, S. Monard, A.S. Perelson, and D.D. Ho. Rapid turnover of T lymphocytes in SIV-infected rhesus macaques. *Science*, 279(5354):1223–1227, 1998.

S.J. Morrison and J. Kimble. Asymmetric and symmetric stem-cell divisions in development and cancer. *Nature*, 441:1068–1074, 2006.

J.L Mull and A. Asakura. A new look at an immortal DNA hypothesis for stem cell self-renewal. *J. Stem Cell Res. Ther.*, 2:e105, 2012.

K. Nakashima and B. de Crombrugghe. Transcriptional mechanisms in osteoblast differentiation and bone formation. *Trends Gen.*, 19:458–466, 2003.

J. Narula, A.M. Smith, B. Gottgens, and O.A. Igoshin. Modeling reveals bistability and low-pass filtering in the network module determining blood stem cell fate. *PLoS Comput. Biol.*, 6(5):e1000771, 2010.

M. Niepel, S.L. Spencer, and P.K. Sorger. Non-genetic cell-to-cell variability and the consequences for pharmacology. *Curr. Opin. Chem. Biol.*, 13(5-6):556–561, 2009.

R.E. Nordon, M. Nakamura, C. Ramirez, and R. Odell. Analysis of growth kinetics by division tracking. *Immunol. Cell. Biol.*, 77(6):523–529, 1999.

S. Palani and C.A. Sarkar. Integrating extrinsic and intrinsic cues into a minimal model of lineage commitment for hematopoietic progenitors. *PLoS Comput. Biol.*, 5(9):e1000518, 2009.

A.M. Parfitt. Osteonal and hemi-osteonal remodeling: The spatial and temporal framework for signal traffic in adult human bone. *J. Cell. Biochem.*, 55:273–286, 1994.

E. Parretta, G. Cassese, A. Santoni, J. Guardiola, A. Vecchio, and F. Di Rosa. Kinetics of in vivo proliferation and death of memory and naive CD8 T cells: Parameter estimation based on 5-Bromo-2'-Deoxyuridine incorporation in spleen, lymph nodes, and bone marrow. *J. Immunol.*, 180(11):7230–7239, 2008.

J.M. Pedraza and J. Paulsson. Effects of molecular memory and bursting on fluctuations in gene expression. *Science*, 319(5861):339–343, 2008.

J. Peltier and D.V. Schaffer. Systems biology approaches to understanding stem cell fate choice. *IET Syst. Biol.*, 4(1):1–11, 2010.

M. Phimphilai, Z. Zhao, H. Boules, H. Roca, and R.T. Franceschi. BMP signaling is required for RUNX2-dependent induction of the osteoblast phenotype. *J. Bone Miner. Res.*, 21(4): 637–646, 2006.

W.A. Prudhomme, K.H. Duggar, and D.A. Lauffenburger. Cell population dynamics model for deconvolution of murine embryonic stem cell self-renewal and differentiation responses to cytokines and extracellular matrix. *Biotechnol. Bioeng.*, 88(3):264–272, 2004.

B.J.C. Quah, H.S. Warren, and C.R. Parish. Monitoring lymphocyte proliferation in vitro and in vivo with the intracellular fluorescent dye carboxyfluorescein diacetate succinimidyl ester. *Nat. Protoc.*, 2(9):2049–2056, 2007.

N. Radde. Fixed point characterization of biological networks with complex graph topology. *Bioinformatics*, 26(22):2874–2880, 2010.

A. Raj and A. van Oudenaarden. Nature, nurture, or chance: stochastic gene expression and its consequences. *Cell*, 135(2):216–226, 2008.

A. Ralston and J. Rossant. Genetic regulation of stem cell origins in the mouse embryo. *Clinical Genetics*, 68(2):106–112, 2005.

D. Ramkrishna. *Population balances – theory and applications to particulate systems in engineering*. Academic Press, 2000.

P. Revy, M. Sospedra, B. Barbour, and A. Trautmann. Functional antigen-independent synapses formed between T cells and dendritic cells. *Nat. Immunol.*, 2:925–931, 2001.

R.M. Ribeiro, H. Mohri, D.D. Ho, and A.S. Perelson. Modeling deuterated glucose labeling of T-lymphocytes. *Bull. Math. Biol.*, 64(2):385–405, 2002.

I. Roeder and I. Glauche. Towards an understanding of lineage specification in hematopoietic stem cells: a mathematical model for the interaction of transcription factors GATA-1 and PU.1. *J. Theor. Biol.*, 241(4):852–865, 2006.

F. Roegiers and Y.N. Jan. Asymmetric cell division. *Curr. Opinion Cell Biol.*, 16(2):195–205, 2004.

H.-M. Ryoo, M.-H. Lee, and Y.-J. Kim. Critical molecular switches involved in BMP-2-induced osteogenic differentiation of mesenchymal cells. *Gene*, 366(1):51–57, 2006.

D. Schittler, J. Hasenauer, F. Allgöwer, and S. Waldherr. Cell differentiation modeled via a coupled two-switch regulatory network. *Chaos*, 20(4):045121, 2010.

D. Schittler, J. Hasenauer, and F. Allgöwer. A generalized population model for cell proliferation: Integrating division numbers and label dynamics. In *Proc. of the 8th Int. Workshop on Computational Systems Biology (WCSB)*, TICSP series # 57, pages 165–168, Zürich, Switzerland, 2011.

D. Schittler, J. Hasenauer, and F. Allgöwer. A model for proliferating cell populations that accounts for cell types. In *Proc. of the 9th Int. Workshop on Computational Systems Biology (WCSB)*, TICSP series # 61, pages 84–87, Ulm, Germany, 2012.

D. Schittler, F. Allgöwer, and R.J. De Boer. A new model to simulate and analyze proliferating cell populations in BrdU labeling experiments. *BMC Syst. Biol.*, 7(Suppl 1):S4, 2013a.

D. Schittler, F. Allgöwer, and S. Waldherr. Multistability equivalence between gene regulatory networks of different dimensionality. In *Proc. of the 12th European Control Conference (ECC)*, pages 3640–3645, Zürich, Switzerland, 2013b.

D. Schittler, T. Jouini, F. Allgöwer, and S. Waldherr. Generalization of the construction method for multistability-equivalent gene regulatory networks to systems with multi-input multi-output loopbreaking. Technical report, arXiv:1312.7250v1 [math.DS], 2013c.

D. Schittler, R.J. De Boer, F. Allgöwer, and J. Hasenauer. A novel model of the BrdU labeling process that mechanistically describes the full label distribution. in prep.

D. Schittler, T. Jouini, F. Allgöwer, and S. Waldherr. Multistability equivalence between gene regulatory networks of different dimensionality with application to a differentiation network. *Int. J. Robust Nonlin.*, submitted.

D. Shemin and D. Rittenberg. The life span of the human red blood cell. *J. Biol. Chem.*, 166: 627–636, 1946.

C.-C. Shu, A. Chatterjee, W.-S. Hu, and D. Ramkrishna. Modeling of gene regulatory processes by population-mediated signaling: New applications of population balances. *Chemical Engineering Science*, pages 188–199, 2012.

C. Shui, T.C. Spelsberg, B.L. Riggs, and S. Khosla. Changes in Runx2/Cbfa1 expression and activity during osteoblastic differentiation of human bone marrow stromal cells. *J. Bone Miner. Res.*, 18(2):213–221, 2003.

F.R. Sidoli, S.P. Asprey, and A. Mantalaris. A coupled single cell-population-balance model for mammalian cell cultures. *Ind. Eng. Chem. Res.*, 45:5801–5811, 2006.

J.-J.E. Slotine and W. Li. *Applied Nonlinear Control*. Prentice-Hall, New Jersey, 1991.

J.A. Smith and L. Martin. Do cells cycle? *Proc. Natl. Acad. Sci. U.S.A.*, 70(4):1263–1267, 1973.

M. Stamatakis. Cell population balance, ensemble and continuum modeling frameworks: Conditional equivalence and hybrid approaches. *Chemical Engineering Science*, 65(2):1008–1015, 2010.

T. Stiehl and A. Marciniak-Czochra. Characterization of stem cells using mathematical models of multistage cell lineages. *Math. Comp. Modelling*, 53(7-8):1505–1517, 2011.

H. Takizawa, R.R. Regoes, C.S. Boddupalli, S. Bonhoeffer, and M.G. Manz. Dynamic variation in cycling of hematopoietic stem cells in steady state and inflammation. *J. Exp. Med.*, 208(2):273–284, 2011.

Q.-Q. Tang, J.-W. Zhang, and M.D. Lane. Sequential gene promoter interactions of C/EBPβ, C/EBPα, and PPARγ during adipogenesis. *Biochem. Biophys. Res. Comm.*, 318(1):235–239, 2004.

R. Thomas. On the relation between the logical structure of systems and their ability to generate multiple steady states or sustained oscillations. In *Numerical methods in the study of critical phenomena*, volume 9 of *Springer Series in Synergetics*, pages 180–193. Springer Berlin Heidelberg, 1981.

R. Thomas. Laws for the dynamics of regulatory networks. *Int. J. Dev. Biol.*, 42:479–485, 1998.

R. Thomas and M. Kaufman. Multistationarity and the basis of cell differentiation and memory. II. logical analysis of regulatory networks in terms of feedback circuits. *Chaos*, 11(1): 180–195, 2001a.

R. Thomas and M. Kaufman. Multistationarity and the basis of cell differentiation and memory. I. structural conditions of multistationarity and other nontrivial behavior. *Chaos*, 11(1): 170–179, 2001b.

W.C. Thompson. *Partial Differential Equation Modeling of Flow Cytometry Data from CFSE-based Proliferation Assays*. Ph.D. thesis, North Carolina State University, U.S.A., 2011.

D.F. Tough and J. Sprent. Lifespan of γ/δ T cells. *J. Exp. Med.*, 187(3):357–365, 1998.

H.M. Tsuchiya, A.G. Fredrickson, and R. Aris. Dynamics of microbial cell populations. *Adv. Chem. Eng.*, 6:125–206, 1966.

A.M. Turing. The chemical basis of morphogenesis. *Phil. Trans. R. Soc. B*, 237(641):37–72, 1952.

J.J. Tyson, K.C. Chen, and B. Novak. Sniffers, buzzers, toggles and blinkers: dynamics of regulatory and signaling pathways in the cell. *Curr. Opin. Cell Biol.*, 15:221–231, 2003.

J.E. Ueckert, G. Nebe von Caron, A.P. Bos, and P.F. Ter Steeg. Flow cytometric analysis of lactobacillus plantarum to monitor lag times, cell divisionand injury. *Letters Appl. Microbiol.*, 25(4):295–299, 1997.

S. Urbani, R. Caporale, L. Lombardini, A. Bosi, and R. Saccardi. Use of CFDA-SE for evaluating the in vitro proliferation pattern of human mesenchymal stem cells. *Cytotherapy*, 8 (3):243–253, 2006.

M. Villani, A. Barbieri, and R. Serra. A dynamical model of genetic networks for cell differentiation. *PLoS One*, 6(3):e17703, 2011.

H. von Foerster. Some remarks on changing populations. In *The kinetics of cellular proliferation*, pages 382–407. Grune and Stratton, New York, 1959.

C.H. Waddington. *Principles of embryology*. George Allen & Unwin Ltd., 1956.

C.H. Waddington. *The strategy of the genes. A discussion of some aspects of theoretical biology*. George Allen & Unwin Ltd., 1957.

S. Waldherr and F. Allgöwer. Searching bifurcations in high-dimensional parameter space via a feedback loop breaking approach. *Int. J. Syst. Sci.*, 40(7):769–782, 2009.

S. Waldherr, S. Streif, and F. Allgöwer. Design of biomolecular network modifications to achieve adaptation. *IET Syst. Biol.*, 6(6):223–231, 2012.

M. Weber, S. Henkel, S. Vlaic, R. Guthke, E. van Zoelen, and D. Driesch. Inference of dynamical gene-regulatory networks based on time-resolved multi-stimuli multi-experiment data applying NetGenerator V2.0. *BMC Syst. Biol.*, 7(1):1, 2013a.

M. Weber, A. Sotoca, P. Kupfer, R. Guthke, and E. van Zoelen. Dynamic modelling of microRNA regulation during mesenchymal stem cell differentiation. *BMC Syst. Biol.*, 7(1): 124, 2013b.

B.A. Welch. The generalization of 'student's' problem when several different population variances are involved. *Biometrika*, 34(1/2):28–35, 1947.

A.D. Wells, H. Gudmundsdottir, and L.A. Turka. Following the fate of individual T cells throughout activation and clonal expansion. Signals from T cell receptor and CD28 differentially regulate the induction and duration of a proliferative response. *J. Clin. Invest.*, 100 (12):3173–83, 1997.

Darren J. Wilkinson. *Stochastic Modelling for Systems Biology*. Chapman & Hall/CRC London, 2006.

A. Wilson, E. Laurenti, G. Oser, R.C. van der Wath, W. Blanco-Bose, M. Jaworski, S. Offner, C.F. Dunant, L. Eshkind, E. Bockamp, P. Lió, H.R. MacDonald, and A. Trumpp. Hematopoietic stem cells reversibly switch from dormancy to self-renewal during homeostasis and repair. *Cell*, 135(6):1118–1129, 2008.

J.A. Winkles, N.L. Tran, S.A.N. Brown, N. Stains, H.E. Cunliffe, and M.E. Berens. Role of Tweak and Fn14 in tumor biology. *Front. Biosci.*, 12:2761–2771, 2007.

D.M. Wittmann, F. Blöchl, D. Trümbach, W. Wurst, N. Prakash, and F.J. Theis. Spatial analysis of expression patterns predicts genetic interactions at the mid-hindbrain boundary. *PLoS Comput. Biol.*, 5(11):e1000569, 2009.

W. Xiong and J.E. Ferrell. A positive-feedback-based bistable 'memory module' that governs a cell fate decision. *Nature*, 426:460–465, 2003.

A. Yates, C. Chan, J. Strid, S. Moon, R. Callard, A. George, and J. Stark. Reconstruction of cell population dynamics using CFSE. *BMC Bioinf.*, 8(1):196, 2007.

G. Zhou, Q. Zheng, F. Engin, E. Munivez, Y. Chen, E. Sebald, D Krakow, and B. Lee. Dominance of SOX9 function over RUNX2 during skeletogenesis. *Proc. Natl. Acad. Sci. U.S.A.*, 103(50):19004–19009, 2006.

J.X. Zhou, L. Brusch, and S. Huang. Predicting pancreas cell fate decisions and reprogramming with a hierarchical multi-attractor model. *PLoS One*, 6(3):e14752, 2011.

G.E. Zinman, S. Zhong, and Z. Bar-Joseph. Biological interaction networks are conserved at the module level. *BMC Syst. Biol.*, 5(1):134, 2011.

M.H. Zwietering, I. Jongenburger, F.M. Rombouts, and K. van 't Riet. Modeling of the bacterial growth curve. *Appl. Environ. Microbiol.*, 56(6):1875–1881, 1990.